Solutions Manual to Accompany

ANALYSIS IN VECTOR SPACES

Solutions Manual to Accompany

ANALYSIS IN VECTOR SPACES

Mustafa A. Akcoglu
University of Toronto
Department of Mathematics
Toronto, Ontario, Canada

Paul F. A. Bartha
University of British Columbia
Department of Philosophy
Vancouver, British Columbia, Canada

Dzung Minh Ha
Ryerson University
Department of Mathematics
Toronto, Ontario, Canada

WILEY

A JOHN WILEY & SONS, INC., PUBLICATION

Published by John Wiley & Sons, Inc., Hoboken, New Jersey.
Published simultaneously in Canada.

ISBN 978-0-470-14825-9

10 9 8 7 6 5 4 3 2 1

CONTENTS

PREFACE

This manual contains solutions to all odd-numbered problems in *Analysis in Vector Spaces*. In a few cases, we have corrected problems stated in the main text. Five problems contained misprints in the main text, which have been corrected in this solutions manual: problems 5.53, 5.55, 9.37, 10.13, and 10.19. In addition, the assumptions stated for two of our problems, Problems 5.39 and 9.31, are incomplete. In both cases, we show here that the conclusions are not correct under the original assumptions, state the complete assumptions, and then solve the revised problems.

PART I

BACKGROUND MATERIAL

CHAPTER 1

SETS AND FUNCTIONS

1.1 SETS IN GENERAL

1.1 Give an example of a family of sets such that any two sets in the family intersect (that is, they have nonempty intersection) but the intersection of all the sets in this family is empty.

Solution. Let $A = \{1, 2, 4\}, B = \{2, 3\}$ and $C = \{3, 4\}$. Then $\{A, B, C\}$ has the desired properties.

1.3 If $\mathcal{A}$ is a collection of subsets of a set X and if $B \subset X$, then show that

$$\begin{aligned}\left(\bigcup_{A\in\mathcal{A}} A\right) \cap B &= \bigcup_{A\in\mathcal{A}} (A \cap B) \text{ and} \\ \left(\bigcap_{A\in\mathcal{A}} A\right) \cup B &= \bigcap_{A\in\mathcal{A}} (A \cup B).\end{aligned}$$

Analysis in Vector Spaces.
By M. A. Akcoglu, P. F. A. Bartha and D. M. Ha

Solution. We have

$$
\begin{aligned}
x \in \left(\bigcup_{A\in\mathcal{A}} A\right) \cap B &\iff x \in \left(\bigcup_{A\in\mathcal{A}} A\right) \text{ and } x \in B \\
&\iff x \in A \text{ for some } A \in \mathcal{A} \text{ and } x \in B \\
&\iff x \in A \cap B \text{ for some } A \in \mathcal{A} \\
&\iff x \in \bigcup_{A\in\mathcal{A}} (A \cap B).
\end{aligned}
$$

Part 2 is proved similarly. Alternatively,

$$
\begin{aligned}
\left(\bigcap_{A\in\mathcal{A}} A\right) \cup B &= \left(\bigcup_{A\in\mathcal{A}} A^c\right)^c \cup B = \left(\left(\bigcup_{A\in\mathcal{A}} A^c\right) \cap B^c\right)^c \\
&= \left(\bigcup_{A\in\mathcal{A}} (A^c \cap B^c)\right)^c = \bigcap_{A\in\mathcal{A}} (A^c \cap B^c)^c \\
&= \bigcap_{A\in\mathcal{A}} (A \cup B)
\end{aligned}
$$

1.5 Show that $A \triangle B \subset (A \triangle C) \cup (C \triangle B)$ for any three sets. Give an example to show that in general the inclusion is a proper inclusion.

Solution. Let us first show that for any sets E, F and G, we have

$$E \setminus F \subset (E \setminus G) \cup (G \setminus F).$$

Let $x \in E \setminus F$. There are two possibilities: either $x \in G$ or $x \notin G$. If $x \in G$, then $x \in G \setminus F$. If $x \notin G$, then $x \in E \setminus G$. Thus,

$$
\begin{aligned}
A \setminus B &\subset (A \setminus C) \cup (C \setminus B) \\
B \setminus A &\subset (B \setminus C) \cup (C \setminus A).
\end{aligned}
$$

Hence,

$$
\begin{aligned}
A \triangle B &= (A \setminus B) \cup (B \setminus A) \\
&\subset ((A \setminus C) \cup (C \setminus B)) \cup ((B \setminus C) \cup (C \setminus A)) \\
&= ((A \setminus C) \cup (C \setminus A)) \cup ((C \setminus B) \cup (B \setminus C)) \\
&= (A \triangle C) \cup (C \triangle B).
\end{aligned}
$$

Let $A = \{1\}$, $B = \{2\}$, and $C = \{3\}$. Then

$$A \triangle B = \{1,\, 2\},\ A \triangle C = \{1,\, 3\},\ C \triangle B = \{2,\, 3\}.$$

In this case $A \triangle B = \{1,\, 2\}$ is a proper subset of $(A \triangle C) \cup (C \triangle B) = \{1,\, 2,\, 3\}$.

1.7 A collection of nonempty subsets of a set X is called a *partition* of X if the sets in this collection are pairwise disjoint and if their union is X. Show that any partition is the family of equivalence classes with respect to an equivalence on X.

Solution. Let $\mathcal{P}$ be a partition of X. Define $R \subset X \times X$ by

$$R = \{ (a, b) \mid \text{there is some } E \in \mathcal{P} \text{ such that } a, b \text{ are in } E \}.$$

Then since the union of sets in $\mathcal{P}$ is X, we see that $(a, a) \in R$ for all $a \in X$. Also, if $(a, b) \in R$, then clearly, $(b, a) \in R$. Finally, let (a, b) and (b, c) be in R. Then there is some $E \in \mathcal{P}$ such that a, b are in E and there is some $F \in \mathcal{P}$ such that b, c are in F. Hence, $b \in E \cap F$. Since the sets in $\mathcal{P}$ are pairwise disjoint. we must have $E = F$. Hence, a, c are in E so that $(a, c) \in R$. Thus, R is an equivalence in X.

Finally, let $F \in \mathcal{P}$. Since $F \neq \emptyset$, there exists some $a \in F$. Then

$$F = \{ x \in X \mid x \in F \} = \{ x \in X \mid (a, x) \in R \},$$

the equivalence class of a under R. Thus, $\mathcal{P} \subset X/R$. Conversely, let E be an equivalence class of some a in X. Since $\mathcal{P}$ is a partition of X, there is a unique $F \in \mathcal{P}$ such that $a \in F$. But of course, $a \in E$. By uniqueness, $E = F$. Consequently, $\mathcal{P} = X/R$.

1.2 SETS OF NUMBERS

1.9 Express the set

$$C = \left\{ x \in \mathbb{R} \;\middle|\; 0 < x^2 - 5x + 4 \le 10 \right\} \subset \mathbb{R}$$

in terms of intervals.

Solution. We have $0 < x^2 - 5x + 4 \le 10$ if and only if

$$(x - 1)(x - 4) > 0 \text{ and } (x - 6)(x + 1) \le 0.$$

Now,

$$\begin{aligned} (x-1)(x-4) > 0 &\iff x \in (-\infty, 1) \cup (4, \infty) \text{ and} \\ (x-6)(x+1) \le 0 &\iff x \in [-1, 6]. \end{aligned}$$

Hence,

$$\begin{aligned} C &= ((-\infty, 1) \cup (4, \infty)) \cap [-1, 6] \\ &= (-\infty, 1) \cap [-1, 6]) \cup ((4, \infty) \cap [-1, 6]) \\ &= [-1, 1) \cup (4, 6]. \end{aligned}$$

1.11 Let $G = \left\{ (x, y) \in \mathbb{R}^2 \mid -1 \le x \le 1, \ -1 \le y \le 1 \right\}$. Consider the sets $U(G, 1)$ and $I(G, 1)$ defined in Example 1.2.7. Express these sets in simpler terms.

Solution. Let $D = [-2, 2] \times [2, 2]$ and set

$$\begin{aligned} R_1 &= ([-2,-1] \times [1,2]) \setminus D_1(-1,1) \\ R_2 &= ([-2,-1] \times [-1,-2]) \setminus D_1(-1,-1) \\ R_3 &= ([1,2] \times [-1,-2]) \setminus D_1(1,-1) \\ R_4 &= ([1,2] \times [1,2]) \setminus D_1(1,1). \end{aligned}$$

Thus, $U(G, 1) = D \setminus (R_1 \cup R_2 \cup R_3 \cup R_4)$.

Since $D_1(-1, 1) \cap D_1(1, -1) = \emptyset$, we have $I(G, 1) = \emptyset$.

1.13 Define a relation $D \subset \mathbb{Z} \times \mathbb{Z}$ as

$$D = \{ (a,\, b) \in \mathbb{Z} \times \mathbb{Z} \mid \text{There is a } k \in \mathbb{Z} \text{ such that } a = kb \} .$$

This relation is called *divisibility*: $(a,\, b) \in D$ just in case a is *divisible by* b. Show that divisibility is reflexive and transitive but not symmetric.

Solution. Since $a = 1a$, we see that $(a, a) \in D$ for all $a \in \mathbb{Z}$. Hence, D is reflexive. If $(a, b) \in D$ and $(b, c) \in D$, then there are integers k, l such that $a = kb$ and $b = lc$. Thus, $a = klc$ and $kl \in \mathbb{Z}$. Hence, $(a, c) \in D$. So, D is transitive. Since $(2, 4) \in D$ but $(4, 2) \notin D$, we see that D is not symmetric.

1.15 Define a relation $C_1 \subset \mathbb{R} \times \mathbb{R}$ by the condition that $(r,\, s) \in C_1$ if and only if $(r - s) \in \mathbb{Z}$. This relation among the real numbers is called *congruence modulo* 1.

1. Show that congruence modulo 1 is an equivalence on $\mathbb{R}$.

2. Show that the interval

$$R = [\,0,\, 1) = \{\, t \in \mathbb{R} \mid 0 \le t < 1 \,\}$$

   is a complete set of representatives for congruence modulo 1.

3. What is the equivalence class represented by $t \in R$?

Note. Let $p \in \mathbb{N}$, $p \geq 2$. Congruence modulo p can be defined on $\mathbb{R}$, but this is not customary.

Solution. (a) We have $(x, x) \in C_1$ for all $x \in \mathbb{R}$ since $x - x = 0 \in \mathbb{Z}$. Let $(x, y) \in C_1$. Then $x - y \in \mathbb{Z}$, so, $y - x \in \mathbb{Z}$. Hence, $(y, x) \in C_1$. Finally, if (x, y) and (y, z) are in C_1, then $x - z = (x - y) + (y - z)$ is in $\mathbb{Z}$ since $x - y \in \mathbb{Z}$ and $y - z \in \mathbb{Z}$. Hence, C_1 is an equivalence relation.

(b) Let $x \in \mathbb{R}$. Then we can find a largest integer n with $n \leq x$. It follows that $x - n \in [0, 1)$ and $(x, n - x) \in C_1$. Hence, the equivalence class E_x of x contains

$x - n$. If r, s are in $[0, 1)$ and r, s are in E_x, then $r - s \in \mathbb{Z}$, and whence, $r = s$. Thus, $[0, 1)$ is a complete set of representatives for $\mathbb{R}$.

(c) If $t \in \mathbb{R}$, then the equivalence class containing t is the set of all numbers of the form $n + t$, where $n \in \mathbb{Z}$.

1.17 Define a relation among the points $(x, y) \in \mathbb{R}^2$ as follows. A point (x, y) is related to a point (x', y') if and only if $xy = x'y'$. Show that this is an equivalence. What are the equivalence classes? What is the equivalence class containing the origin $(0, 0)$? What is a complete set of representatives? Is the line

$$\{ (x, y) \in \mathbb{R}^2 \mid x = y \}$$

a complete set of representatives? Why or why not?

Solution. Let $\sim$ denote the relation in the question. It is clear that $(a, b) \sim (a, b)$ for all $(a, b) \in \mathbb{R}^2$. Also, if $(a, b) \sim (c, d)$, then $(c, d) \sim (a, b)$. Finally, if $(a, b) \sim (c, d)$ and $(c, d) \sim (e, f)$, then $ab = cd$ and $cd = ef$, so that $ab = ef$. Hence, $(a, b) \sim (e, f)$. Thus, $\sim$ is an equivalence relation in $\mathbb{R}^2$.

The equivalence classes induced by $\sim$ are the sets $\{ (x, y) \mid xy = r \}$, where $r \in \mathbb{R}$. If $r \neq 0$, then these equivalence classes are simply the hyperbolas.

The equivalence class containing $(0, 0)$ is the set $\{ (x, y) \mid x = 0 \text{ or } y = 0 \}$. Thus, it consists of the 2 axes.

Let $S = \{ (x, y) \in \mathbb{R}^2 \mid x = y \}$. Then S is not a complete set of representatives because the equivalence class of $(-2, 1)$ does not contain any member of S.

1.19 Let $n \in \mathbb{N}$. Show by induction that $4^n - 3n - 1$ is divisible by 9. (Divisibility is defined in Problem 1.2.)

Solution. Let S be the set of all positive integers n such that $4^n - 3n - 1$ is divisible by 9. We want to show that $S = \mathbb{N}$. To apply the induction principle, we need to verify that $1 \in S$ and whenever $k \in S$, then $k + 1$ also belongs to S.

It is clear that $1 \in S$ since $4^1 - 3 - 1 = 0$ is divisible by 9. Suppose that $k \in S$. Then there is some integer d such that $4^k - 3k - 1 = 9d$. Thus,

$$\begin{aligned} 4^{k+1} - 3(k + 1) - 1 &= 4(4^k) - 3k - 1 - 3 = 4^k - 3k - 1 + 3(4^k) - 3 \\ &= 9d - 3(4^k - 1). \end{aligned}$$

Remember that for any positive integer k and any real number a, we have $a^k - 1 = (a - 1)(a^{k-1} + a^{k-2} + \cdots + 1)$. Thus, $4^k - 1 = 3(4^{k-1} + \cdots + 1)$. Hence,

$$4^{k+1} - 3(k + 1) - 1 = 9d - 3(4^k - 1) = 9d - 9(4^{k-1} + \cdots + 1) = 9\ell,$$

where $\ell = d - (4^{k-1} + \cdots + 1)$. Since both d and $4^{k-1} + \cdots + 1$ are integers, so is ℓ. Hence, $4^{k+1} - 3(k + 1) - 1$ is divisible by 9, so that, $k + 1 \in S$.

1.21 (*Binomial Theorem*) Let $a,\ b \in \mathbb{R}$ and $n \in \mathbb{Z}^+$. Use the induction principle to show that

$$(a+b)^n = \sum_{k=0}^{n} \binom{n}{k} a^{n-k} b^k .$$

Solution. The assertion is true for $n = 1$ because

$$\sum_{k=0}^{1} \binom{1}{k} a^{1-k} b^k = \binom{1}{0} ab^0 + \binom{1}{1} a^0 b^1 = a + b = (a+b)^1 .$$

Assume that for some $n \in \mathbb{Z}^+$, we have

$$(a+b)^n = \sum_{k=0}^{n} \binom{n}{k} a^{n-k} b^k .$$

Then

$$\begin{aligned}
(a+b)^{n+1} &= (a+b)(a+b)^n \\
&= (a+b)\sum_{k=0}^{n} \binom{n}{k} a^{n-k} b^k \\
&= \sum_{k=0}^{n} \binom{n}{k} a^{n-k+1} b^k + \sum_{k=0}^{n} \binom{n}{k} a^{n-k} b^{k+1} \\
&= \sum_{k=0}^{n} \binom{n}{k} a^{n-k+1} b^k + \sum_{k=1}^{n+1} \binom{n}{k-1} a^{n-(k-1)} b^k \\
&= \left(\sum_{k=1}^{n} \binom{n}{k} a^{n-k+1} b^k + \binom{n}{0} a^{n+1} b^0\right) \\
&\quad + \left(\sum_{k=1}^{n} \binom{n}{k-1} a^{n-(k-1)} b^k + \binom{n}{n} a^0 b^{n+1}\right) \\
&= \sum_{k=1}^{n} \left[\binom{n}{k} + \binom{n}{k-1}\right] a^{n+1-k} b^k \\
&\quad + \binom{n}{0} a^{n+1} b^0 + \binom{n}{n} a^0 b^{n+1} \\
&= \sum_{k=1}^{n} \left[\binom{n}{k} + \binom{n}{k-1}\right] a^{n+1-k} b^k + a^{n+1} b^0 + a^0 b^{n+1} \\
&= \sum_{k=1}^{n} \binom{n+1}{k} a^{n+1-k} b^k + a^{n+1} + b^{n+1} \\
&= \sum_{k=0}^{n+1} \binom{n+1}{k} a^{n+1-k} b^k .
\end{aligned}$$

1.23 Let $n \in \mathbb{N}$. Show that $\sum_{k=1}^{n} k^2 = (1/6)n(n+1)(2n+1)$.

Solution. We can give a solution based on induction, however, the solution to be given below also allows us to discover formulas for $\sum_{k=1}^{n} k^m$, where m is a given positive integer.

Let $S = \sum_{k=1}^{n} k^2$. We have $(k+1)^3 - k^3 = 3k^2 + 3k + 1$ so that

$$\begin{aligned}\sum_{k=1}^{n} [(k+1)^3 - k^3] &= 3\sum_{k=1}^{n} k^2 + 3\sum_{k=1}^{n} k + \sum_{k=1}^{n} 1 \\ &= 3S + 3\frac{n(n+1)}{2} + n.\end{aligned}$$

On the other hand,

$$\begin{aligned}\sum_{k=1}^{n} [(k+1)^3 - k^3] &= (2^3 - 1^3) + (3^3 - 2^3) + \cdots + [(n+1)^3 - n^3] \\ &= (n+1)^3 - 1^3 = n^3 + 3n^2 + 3n.\end{aligned}$$

Hence, $n^3 + 3n^2 + 3n = 3S + n + 3n(n+1)/2$ so that

$$3S = n^3 + 3n^2 + 2n - 3\frac{n(n+1)}{2}.$$

After some simplifications, we obtain $S = (1/6)n(n+1)(2n+1)$.

1.3 FUNCTIONS

1.25 Let $f : D \to Y$ be a function with the range space Y. Let $\mathcal{F}$ be a collection of subsets of Y. Show that

$$f^{-1}\left(\bigcup_{F\in\mathcal{F}} F\right) = \bigcup_{F\in\mathcal{F}} f^{-1}(F) \text{ and } f^{-1}\left(\bigcap_{F\in\mathcal{F}} F\right) = \bigcap_{F\in\mathcal{F}} f^{-1}(F).$$

Solution. Let $x \in D$. We have

$$\begin{aligned}x \in f^{-1}\left(\bigcup_{F\in\mathcal{F}} F\right) &\iff f(x) \in \bigcup_{F\in\mathcal{F}} F \\ &\iff f(x) \in F \quad \text{for some } F \in \mathcal{F} \\ &\iff x \in f^{-1}(F) \quad \text{for some } F \in \mathcal{F} \\ &\iff x \in \bigcup_{F\in\mathcal{F}} f^{-1}(F).\end{aligned}$$

Similarly,

$$\begin{aligned}x \in f^{-1}\left(\bigcap_{F\in\mathcal{F}} F\right) &\iff f(x) \in \bigcap_{F\in\mathcal{F}} F \\ &\iff f(x) \in F \quad \text{for all } F \in \mathcal{F} \\ &\iff x \in f^{-1}(F) \quad \text{for all } F \in \mathcal{F} \\ &\iff x \in \bigcap_{F\in\mathcal{F}} f^{-1}(F).\end{aligned}$$

1.27 Let $f(x) = x^2 - 6x - 7$. What is the range $f(D)$ of f? Is f one-to-one on its domain? If not, find two different sets P, $Q \subset \mathbb{R}$ such that f is one-to-one on P and one-to-one on Q and such that $f(P) = f(Q) = f(\mathbb{R})$. Find the inverse function of f on P and the inverse function of f on Q.

Solution. We have $f(x) = (x-3)^2 - 16$ for all $x \in \mathbb{R}$. Hence,

$$f(\mathbb{R}) = [-16, \infty).$$

Since $f(2) = f(4)$, it is not one-to-one.

Let $P = (-\infty, 3], Q = [3, \infty)$. Then f is one-to-one on both P and on Q, and $f(P) = f(Q) = f(\mathbb{R})$.

Suppose that $f(x) = y$. Then $y + 16 = (x-3)^2$ so that

$$x = 3 \pm \sqrt{y + 16}.$$

It follows that the inverse of f on P is given by

$$g(y) = 3 - \sqrt{y + 16} \quad \text{for all } y \in P.$$

Similarly, the inverse of f on Q is

$$h(y) = 3 + \sqrt{y + 16} \quad \text{for all } y \in Q.$$

1.29 Let $f : \mathbb{R}^2 \to \mathbb{R}^2$ be the polar coordinates defined by

$$(x, y) = f(r, \theta) = (r\cos\theta, r\sin\theta).$$

Find $f^{-1}(A)$ for

$$A = \{ (x, y) \in \mathbb{R}^2 \mid 1 \le (x^2 + y^2) \le 4 \text{ and } 0 \le (y/x) \le 1 \}.$$

Solution. Let $(r, \theta) \in f^{-1}(A)$. Then $(r\cos\theta, r\sin\theta) \in A$ so that

$$1 \le r^2(\cos^2\theta + \sin^2\theta) \le 4 \quad \text{and} \quad 0 \le \frac{r\sin\theta}{r\cos\theta} \le 1.$$

Hence, $1 \le |r| \le 2$ and $k\pi \le \theta \le \pi/4 + k\pi$ for some $k \in \mathbb{Z}$. That is,

$$f^{-1}(A) = \{ (r, \theta) \mid 1 \le |r| \le 2, k\pi \le \theta \le \pi/4 + k\pi \text{ for some } k \in \mathbb{Z} \}.$$

1.31 Define the function f from the xy-plane to the uv-plane by

$$(u, v) = f(x, y) = (3x + 2y, 6x - 4y).$$

Repeat the parts of Problem 1.30 for this example.

Solution. (a) Domain of f is $\mathbb{R}^2$.

(b) Let $(a, b) \in \mathbb{R}^2$. Then the system

$$\begin{aligned} 3x + 2y &= a \\ 6x - 4y &= b \end{aligned}$$

has a unique solution $(x, y) \in \mathbb{R}^2$. We obtain

$$(x,\, y) = ((2a + b)/12,\, (2a - b)/4)$$

and $f((2a + b)/12,\, (2a - b)/4) = (a, b)$. Hence the range of f is $\mathbb{R}^2$.

(c) We have $(x, y) \in f^{-1}(L_a) \iff 3x + 2y = a \iff y = (a - 3x)/2$. Thus,

$$f^{-1}(L_a) = \left\{ \left(x,\, \frac{a - 3x}{2}\right) \mid x \in \mathbb{R} \right\}.$$

Similarly, $(x, y) \in f^{-1}(M_b) \iff 6x - 4y = b \iff y = (6x - b)/4$. Thus,

$$f^{-1}(M_b) = \left\{ \left(x,\, \frac{6x - b}{4}\right) \mid x \in \mathbb{R} \right\}.$$

(d) From part (b), we have

$$f^{-1}(\{(a, b)\}) = \left\{ \left(\frac{2a + b}{12},\, \frac{2a - b}{4}\right) \right\}.$$

(Alternatively, $f^{-1}(\{(a, b)\}) = f^{-1}(L_a \cap M_b) = f^{-1}(L_a) \cap f^{-1}(M_b)$ and so on.)

(e) From our solution in part (d), we see that f is one-to-one on all of $\mathbb{R}^2$.

1.33 Define the function f from the xy-plane to the uv-plane by

$$(u,\, v) = f(x,\, y) = ((x^2 + y^2)/(2x),\, (x^2 + y^2)/(2y)).$$

Repeat the parts of Problem 1.30 for this example. (*Hint.* In Part A of Figure 1.4 we see the inverse images of the lines $u = -3,\, 3,\, 5,\, 7$ and the lines $v = -3,\, 3,\, 5,\, 7$.)

Solution. This function is defined if and only if both x and y are nonzero. Hence the domain of f is $D = \{\, (x,\, y) \mid xy \neq 0 \,\}$. We claim that f is one-to-one on D and that $R = \{\, (u,\, v) \mid uv \neq 0 \,\}$ is the range of f. In fact, if $(u,\, v) \in R$ then we see that

$$x = 2uv^2/(u^2 + v^2) \text{ and } y = 2u^2v/(u^2 + v^2)$$

is the unique point in D with $f(x, y) = (u, v)$. Also, the inverse image of the $u = a$ line is the circle $(x - a)^2 + y^2 = a^2$ and the inverse image of the $v = b$ line is the circle $x^2 + (y - b)^2 = b^2$. Any two such circles intersect only at one point (x, y) with nonzero coordinates (that is $xy \neq 0$). Since f is one-to-one on D we see that $f(A) = f(D)$ only if $A = D$.

1.35 Define the function f from the xy-plane to the uv-plane by

$$(u, v) = f(x, y) = (p(x, y) + q(x, y), p(x, y) - q(x, y)),$$

where $p(x, y) = ((x + 1)^2 + y^2)^{1/2}$ and $q(x, y) = ((x - 1)^2 + y^2)^{1/2}$. Repeat the parts of Problem 1.30 for this example.

Solution. This function is defined for all $(x, y) \in \mathbb{R}^2$. Hence $D = \mathbb{R}^2$. We claim that the range R of f is the region

$$S = \{ (u, v) \mid 2 \leq u \text{ and } -2 \leq v \leq 2 \}.$$

To see that $R \subset S$ note that $p(x, y)$ is the distance of (x, y) to $(-1, 0)$ and $q(x, y)$ is the distance of (x, y) to $(1, 0)$. There are two important inequalities involving the sum and the difference of these distances. In general terms, let A, B, and P be three points in the plane. Then $\overline{PA} + \overline{PB} \geq \overline{AB}$ and $|\overline{PA} - \overline{PB}| \leq \overline{AB}$. They can be verified by simple analytic geometry or otherwise. (See Remarks 4.1.8 for more general forms of these inequalities.) This shows that $R \subset S$.

Now we show that that $S \subset R$. Consider the equation $u(x, y) = 2a$ with $1 \leq a$. By elementary algebraic manipulation we see that if $a > 1$ then this is the equation of the ellipse $(x^2/a^2) + (y^2/b^2) = 1$ where $b = (a^2 - 1)^{1/2}$. If $a = 1$ this equation reduces to $y = 0$ and $|x| \leq 1$, which is a degenerate ellipse. Given any point in xy-plane there is exactly one of these ellipses that passes through this point, corresponding to all values of $1 \leq a$. Similarly, consider the equation $v(x, y) = 2\alpha$ with $|\alpha| \leq 1$. If $0 < |\alpha| < 1$ then this is the equation of a branch of the hyperbola $(x/\alpha)^2 - (y/\beta)^2 = 1$ where $\beta = (1 - \alpha^2)^{1/2}$. If $0\alpha < 1$ then this branch is the right-hand branch and if $-1 < \alpha < 0$ then it is the left-hand branch. Also, $v(x, y) = 0$ is the equation of the y-axis, $v(x, y) = 1$ is equation of the segment $(x, 0)$, $1 \leq x$, and $v(x, y) = -1$ is equation of the segment $(x, 0)$, $x \leq -1$. All these three lines are degenerate hyperbola branches. Given any point in xy-plane there is exactly one of these hyperbola branches that passes through this point, corresponding to all values of $|a| \leq 1$. This shows that $S = R$.

Also, we verify that a non-degenerate ellipse cuts the y-axis or a non-degenerate hyperbola branch at exactly two points, symmetrical with respect to the x-axis. Hence we see that f is one-to-one on the upper half-plane $A = \{ (x, y) \mid y \geq 0 \}$. Also, $f(A) = f(\mathbb{R}^2) = f(D) = S$.

CHAPTER 2

REAL NUMBERS

2.1 REVIEW OF THE ORDER RELATIONS

2.1 Show that if $0 < a < b$, then $0 < (1/b) < (1/a)$. Use only the two properties of positive numbers listed in Remark 2.1.1.

Solution. First show that if $0 < a$ then $0 < (1/a)$. Now $(1/a) \neq 0$, since $a(1/a) = 1$. If $(1/a) \notin P$ then $-(1/a) \in P$. Hence

$$-(1/a)a = -1 \in P.$$

This is a contradiction. Hence $(1/a) \in P$. Also $(1/b) \in P$ and

$$(1/a)(1/b) \in P.$$

Then

$$(1/a) - (1/b) = (b - a)(1/a)(1/b) \in P.$$

Analysis in Vector Spaces.
By M. A. Akcoglu, P. F. A. Bartha and D. M. Ha

Hence $(1/b) < (1/a)$.

2.3 Let S be the set of all points $(x,\ y)$ in the xy-plane such that

$$|x-2|+|y-3| \leq 1.$$

Show that S is a region bounded by four lines. Find these lines and describe S geometrically.

Solution. Consider the region R in the uv-plane defined by $|u|+|v| \leq 1$. Consider the four quarters of the uv-plane separately.

If $u \geq 0,\ v \geq 0$ then $(u+v) \leq 1$.

If $u \leq 0,\ v \leq 0$ then $-(u+v) \leq 1$ or $-1 \leq (u+v)$.

If $u \geq 0,\ v \leq 0$ then $(u-v) \leq 1$.

If $u \leq 0,\ v \geq 0$ then $(-u+v) \leq 1$ or $-1 \leq (u-v)$.

Hence R is the square bounded by the lines $(u+v) = \pm 1$ and $(u-v) = \pm 1$.

To obtain S in the xy-plane we let $u = x-2$ and $v = y-3$. We see that S is the square centered at the point $(2,\ 3)$ and bounded by the lines $x+y=4$, $x+y=6$, $x-y=0$, and $x-y=2$.

2.5 Let a, b, and c be real numbers with $a > 0$. Show that there is an $M \in \mathbb{R}$ such that if $|x| > M$, then $ax^2+bx+c > 0$.

Solution. We have

$$ax^2+bx+c = a\left(x+\frac{b}{2a}\right)^2 + \frac{4ac-b^2}{4a}.$$

Thus,

$$ax^2+bx+c > 0 \Longleftrightarrow \left(x+\frac{b}{2a}\right)^2 > \frac{b^2-4ac}{4a^2}.$$

We may assume that $b^2-4ac > 0$ (or else any M will suffice). Then

$$\begin{aligned} ax^2+bx+c>0 \quad &\Longleftrightarrow \quad \left|x+\frac{b}{2a}\right| > \frac{\sqrt{b^2-4ac}}{2a} \\ &\Longleftrightarrow \quad x > \frac{-b+\sqrt{b^2-4ac}}{2a} \quad \text{or} \quad x < \frac{-b-\sqrt{b^2-4ac}}{2a}. \end{aligned}$$

Take M to be the larger of the two numbers $\left|\frac{-b+\sqrt{b^2-4ac}}{2a}\right|$ and $\left|\frac{b+\sqrt{b^2-4ac}}{2a}\right|$.

2.7 Let a, b, and c be real numbers. Show that

$$a^2 + b^2 + c^2 \geq ab + bc + ca.$$

Solution. Let a, b, c be real numbers. Then

$$2(a^2 + b^2 + c^2 - ab - bc - ca) = (a-b)^2 + (b-c)^2 + (c-a)^2 \geq 0.$$

2.2 COMPLETENESS OF REAL NUMBERS

2.9 If $a \in \mathbb{R}$ and $S \subset \mathbb{R}$, then let aS denote the set of all real numbers of the form $x = as$, where $s \in S$. Show that if $\sup S$ exists and if $a > 0$, then $\sup(aS) = a \sup S$. If $a < 0$, then show that $\inf(aS) = a \sup S$.

Solution. Let $a > 0$ and $S' = aS$. Let $M = \sup S$. Hence $s \leq M$ for all $s \in S$ implies $as \leq aM$ for all $s \in S$. Hence aM is an upper bound for $S' = aS$. Therefore $M' = \sup S'$ exists and $M' \leq aM$. But $S = (1/a)S'$ and $(1/a) > 0$. Hence by the same reasoning we see that $M \leq (1/a)M'$ or that $aM \leq M'$. Hence $M' = aM$.

Now let $a < 0$ and $S' = aS$ Let $M = \sup S$. Hence $s \leq M$ for all $s \in S$ implies $as \geq aM$ for all $s \in S$. Hence aM is a lower bound for $S' = aS$. Therefore $m' = \inf S'$ exists and $m' \geq aM$. But $m' \leq as$ for all $s \in S$ implies that $(1/a)m' \geq s$ for all $s \in S$. Therefore $(1/a)m'$ is an upper bound for S and $(1/a)m' \geq M$. Hence $m' \leq aM$. Therefore $m' = aM$.

2.11 Let $E \subset \mathbb{R}$ be nonempty and bounded. Show that $\mathbb{R} \setminus E$ is bounded neither from above nor below.

Solution. Let $E \subset \mathbb{R}$ be nonempty and bounded. Then there are real numbers $a < b$ such that $E \subset [a, b]$. Hence,

$$\mathbb{R} \setminus E \supset R \setminus [a, b] = (-\infty, a) \cup (b, \infty).$$

Since $(-\infty, a) \cup (b, \infty)$ is neither bounded from below nor above, the same is true of the larger set $\mathbb{R} \setminus E$.

2.13 Let a, b, c, and d be in $\mathbb{R}$ with $a < b$ and $c < d$. Set

$$T = \{\, |2x - 3y| \mid x \in [a, b], y \in [c, d] \,\}.$$

Show that T is bounded. Find $\sup T$ and $\inf T$.

Solution. Let $x \in [a,b], y \in [c,d]$. Then $2x \leq 2b$ and $3y \geq 3c$. Hence, $2x - 3y \leq 2b - 3c$. Also, $x \geq a$ and $y \leq d$ so that $2x - 3y \geq 2a - 3d$. Hence,

$$2a - 3d \leq 2x - 3y \leq 2b - 3c.$$

It follows easily that $|2x - 3y| \leq M$ where $M = \max\{|2a - 3d|, |2b - 3c|\}$. Hence, S is bounded.

The infimum and supremum of S can be computed as follows:

If $2a - 3d > 0$, then $2a - 3d \leq |2x - y| \leq 2b - 3c$ so that

$$\inf S = 2a - 3d, \ \sup S = 2a - 3c.$$

If $2b - 3c < 0$, then $3c - 2b \leq |2x - 3y| \leq 3d - 2a$ so that

$$\inf S = 3c - 2b, \ \sup S = 3d - 2a.$$

If $2a - 3d < 0$ and $2b - 3c > 0$, then $\sup S = 2b - 3c$. Also, in this case, $2x - 3y = 0$ for some $x \in [a,b]$ and some $y \in [c,d]$. Hence, $\inf S = 0$.

2.3 SEQUENCES OF REAL NUMBERS

2.15 Let $|a| < 1$. Show that $\lim_n a^n = 0$.

Solution. Let $u_n = a^n$. Since $-1 < a < 1$, the sequence $(|u_n|)$ is bounded. Also, $|u_{n+1}| = |a^{n+1}| = |a||a^n| = |a||u_n| < |u_n|$, we see that $(|u_n|)$ is monotone. Hence, $\lim |u_n|$ exists. Let $l = \lim |u_n|$. Then

$$l = \lim |u_{n+1}| = \lim |a||u_n| = |a| \lim |u_n| = |a|l.$$

If $l \neq 0$, then $|a| = 1$, a contradiction. Hence, we must have $l = 0$. Since $\lim |u_n| = 0$, it follows that $\lim u_n = 0$ also.

2.17 Show that each irrational number is the limit of a sequence of rational numbers. Also show that any rational number is the limit of a sequence of irrational numbers.

Solution. Let x be an irrational number. For each $n \in \mathbb{N}$, there is some rational number r_n in $[x, x + 1/n)$. Thus, (r_n) is a sequence of rational numbers such that $|x - r_n| < 1/n$ for all n. Hence, $r_n \to x$. Similarly, let r be a rational number. For each $n \in \mathbb{N}$, there is some irrational number u_n in $[r, r + 1/n)$. Thus, (u_n) is a sequence of irrational numbers such that $|r - u_n| < 1/n$ for all $n \in \mathbb{N}$. Hence, $u_n \to r$.

2.19 Let $|a| < 1$ and $r \in \mathbb{R}$. Show that the sequence

$$x_n = \sum_{k=1}^{n} \binom{r}{k} a^k$$

is convergent. (The limit of this sequence is $(1+a)^r$, but the proof of this fact requires more work.)

Solution. We see that

$$\begin{aligned} |x_{n+1} - x_n| &= \binom{r}{n+1} a^{n+1} = a\frac{r-n}{n+1}\binom{r}{n} a^n \\ &= a\frac{r-n}{n+1}\,|x_n - x_{n-1}|\,. \end{aligned}$$

Let $b_n = |a(r-n)|\,(n+1)^{-1}$. Then $\lim_n b_n = |a| < 1$. Hence there is a $c < 1$ and an $n \in \mathbb{N}$ such that $b_n < c$ for all $n \geq N$. Then Example 2.3.15 shows that that x_n is convergent.

2.21 Let $a > 0$ and $c > 0$. Define a sequence x_n recursively by $x_1 = a$ and

$$x_{n+1} = 1 + c(1+x_n)^{-1},\ n \in \mathbb{N}.$$

Show that x_n is a convergent sequence. What is $\lim_n x_n$?

Solution. Note that $1 \leq x_n$ for all $n \in \mathbb{N}$. Also

$$\begin{aligned} (1+x_n)(1+x_{n-1}) &= (2 + c(1+x_{n-1})^{-1})(1+x_{n-1}) \\ &= (2(1+x_{n-1}) + c) \geq (4+c). \end{aligned}$$

Hence

$$|x_{n+1} - x_n| = \frac{c\,|x_n - x_{n-1}|}{(1+x_n)(1+x_{n-1})} \leq \frac{c}{c+4}\,|x_n - x_{n-1}|.$$

Then x_n converges, since $c(4+c)^{-1} < 1$. If $x_n \to a$ then $a = 1 + c(1+a)^{-1}$ and $a^2 = c+1$. Since $a \geq 0$ we obtain $a = \sqrt{1+c}$.

2.4 SUBSEQUENCES

2.23 Let x_n be a sequence in $\mathbb{R}$. Let $a \in \mathbb{R}$. Assume that

$$\mathbb{L}_r = \{\, n \in \mathbb{N} \mid a - r < x_n \leq a \,\}$$

is an infinite (unbounded) set of integers for each $r > 0$. Show that x_n has a monotone increasing subsequence converging to a.

Solution. Distinguish two cases. If the set

$$\mathbb{K} = \{\, n \in \mathbb{N} \mid x_n = a \,\} \subset \mathbb{N}$$

is an infinite set then the constant subsequence x_k, $k \in \mathbb{K}$, is a monotone increasing sequence converging to a.

Now assume that $\mathbb{K}$ is a finite set. Let

$$\mathbb{L}_r' = \{ n \in \mathbb{N} \mid a - r < x_n < a \} \subset \mathbb{N}.$$

Then $\mathbb{L}_r'$ is an infinite set for each $r > 0$, since $\mathbb{L}_r = \mathbb{L}_r' \cup \mathbb{K}$. Choose a $k_1 \in \mathbb{L}_1'$. Assume that $k_1, \ldots, k_{n-1}$ are already chosen and satisfy

$$x_{k_1} < \cdots < x_{k_{n-1}} < a.$$

Let r_n be any number such that $0 < r_n < (a - x_{k_{n-1}})$ and $r_n < (1/n)$. Choose $k_n \in \mathbb{L}_{r_n}'$. Then, clearly,

$$x_{k_n} < x_{k_{n+1}} < a \text{ and } (a - (1/n)) < x_{k_n} < a.$$

Hence x_{k_n} is a monotone increasing sequence converging to a.

Note that here we use only that $\mathbb{L}_r'$ is nonempty for each $r > 0$. But this is equivalent to the the fact that $\mathbb{L}_r'$ is infinite for each $r > 0$.

2.25 Let x_n be a bounded sequence in $\mathbb{R}$. For each $n \in \mathbb{N}$, let

$$S_n = \{ x \in \mathbb{R} \mid x = x_k,\ n \leq k \}.$$

Then S_n is a nonempty and bounded set and $s_n = \sup S_n$ exists. Show that s_n is a convergent sequence. Show that if $s_n \to a$, then x_n has a subsequence converging to a. (This gives another proof for the Bolzano-Weierstrass theorem.)

Solution. If $\emptyset \neq S' \subset S \subset \mathbb{R}$ and if $M = \sup S$ exists then M is obviously an upper bound for S'. Therefore $\sup S' = M'$ also exists and $M' \leq M$. Also, if $m = \inf S$ exists then $m' = \inf S'$ also exists and $m \leq m'$.

We see that $\emptyset \neq S_{n+1} \subset S_n$ for all $n \in \mathbb{N}$. Since $s_1 = \sup S_1$ exists all $s_n = \sup S_n$ also exist and $s_{n+1} \leq s_n$ for all $n \in \mathbb{N}$. Therefore s_n is a monotone decreasing sequence and bounded below by $\inf S_1$. Therefore $a = \lim_n s_n$ exists by the monotone convergence theorem.

We claim that given any $\varepsilon > 0$ and any $N \in \mathbb{N}$ there is a $k \geq N$ such that $|x_k - a| < \varepsilon$. In fact first find an $n \geq N$ such that $a \leq s_n < a + \varepsilon$. This is possible since $a = \inf_n s_n$ is the limit of the monotone decreasing sequence s_n. Then there must be a $k \geq n$ such that $x_k \in (s_n - \varepsilon,\ s_n] \subset (a - \varepsilon,\ a + \varepsilon)$, for otherwise s_n would not be the least upper bound of S_n.

Let $I_n = (a - (1/n),\ a + (1/n))$ for each $n \in \mathbb{N}$. To obtain a subsequence x_{k_n} converging to a find $x_{k_1} \in I_1$. Assume that k_n is obtained such that $x_{k_n} \in I_n$. Then choose $k_{n+1} > k_n$ such that $x_{k_{n+1}} \in I_{n+1}$. We see that the subsequence x_{k_n} converges to a.

2.27 Let x_n be a bounded sequence in $\mathbb{R}$. Let s_n be the sequence obtained in Problem 2.4. Assume that there is no $N \in \mathbb{N}$ such that $s_N = s_{N+k}$ for all $k \in \mathbb{N}$. Show that x_n has a monotone decreasing convergent subsequence.

Solution. Let $a = \lim_n s_n$. Then $a \neq s_n$ for each n. Hence for all $r > 0$ there is an s_n such that $a < s_n < a + r$. Then there is also an $k \geq n$ such that $a < x_k \leq s_n$. This means that

$$\mathbb{S}_r = \{ n \in \mathbb{N} \mid a \leq x_n < a + r \} \subset \mathbb{N}$$

is an infinite set of integers for each $r > 0$. Then Problem 2.25 shows that x_n has a monotone decreasing subsequence converging to a.

2.29 Show that a sequence x_n is a Cauchy sequence if and only if there is a zero sequence z_n such that $|x_n - x_m| \leq z_n$ for all $m,\, n \in \mathbb{N}$, $m \geq n$.

Solution. Let x_n be a Cauchy sequence. Then x_n is a bounded sequence. For each $n \in \mathbb{N}$ let S_n be the set of numbers $|x_n - x_m|$ with $m \geq n$. Hence S_n is a bounded set and $z_n = \sup S_n$ exists. Then we have $|x_n - x_m| \leq z_n$ for all $m \geq n$. Now given $\varepsilon > 0$ choose an $N \in \mathbb{N}$ such that $|x_m - x_n| \leq \varepsilon$ for all $m \geq N$ and $n \geq N$. This means that $z_n \leq \varepsilon$ for all $n \geq N$. Hence z_n is a zero sequence.

Conversely assume that there is a zero sequence z_n such that $|x_n - x_m| \leq z_n$ for all $n \in \mathbb{N}$ and for all $m \geq n$. Given $\varepsilon > 0$ find an $N \in \mathbb{N}$ such that $z_N \leq \varepsilon/2$. If $m,\, n \geq N$ then we see that

$$|x_N - x_n| \leq (\varepsilon/2) \text{ and } |x_N - x_m| \leq (\varepsilon/2).$$

This gives $|x_n - x_m| \leq \varepsilon$ for all $m,\, n \geq N$. Hence x_n is a Cauchy sequence.

2.31 Give an example of a nonconvergent sequence x_n such that

$$|x_{n+2} - x_{n+1}| < |x_{n+1} - x_n| \text{ for all } n \in \mathbb{N}.$$

Solution. Define x_n inductively as $x_1 = 1$ and $x_{n+1} = x_n + 1 + (1/n)$. This is obviously not a Cauchy sequence, since $(x_{n+1} - x_n) > 1$ for all $n \in \mathbb{N}$. But

$$|x_{n+2} - x_{n+1}| = 1 + (n+1)^{-1} < 1 + n^{-1} = |x_{n+1} - x_n|$$

for all $n \in \mathbb{N}$.

2.33 Give an example of a Cauchy sequence x_n for which the sequence

$$s_n = |x_2 - x_1| + |x_3 - x_2| + \cdots + |x_{n+1} - x_n|\,,\; n \in \mathbb{N},$$

is an unbounded sequence.

Solution. There are sequences of positive numbers a_n converging to zero such that $r_n = a_1 + \cdots + a_n$ is unbounded. Take for example the sequence

$$1,\ (1/2),\ (1/2),\ (1/3),\ (1/3),\ (1/3),\ (1/4),\ (1/4),\ (1/4),\ (1/4),\ \ldots.$$

It consists of consecutive segments. In the n th segment the number $(1/n)$ is repeated n times. We see that the sum over each segment is 1. Hence the total sums are unbounded.

Let a_n be a sequence as specified above. Let $x_n = (-1)^n a_n$. Then x_n still converges to zero. Hence it is a Cauchy sequence. But

$$|x_{n+1} - x_n| = a_{n+1} + a_n$$

for each $n \in \mathbb{N}$. In fact, a_i s are always nonnegative and x_{n+1} and x_n have always the opposite signs. Hence

$$s_n \geq r_n = a_1 + \cdots + a_n$$

is an unbounded sequence.

2.35 Show that every Cauchy sequence x_n has a subsequence x_{k_n} for which

$$s_n = |x_{k_2} - x_{k_1}| + |x_{k_3} - x_{k_2}| + \cdots + |x_{k_{n+1}} - x_{k_n}|,\ n \in \mathbb{N},$$

is a bounded sequence.

Solution. Let p_n be a sequence of positive numbers such that

$$r_n = p_1 + \cdots + p_n$$

is a bounded sequence. Let for example $p_n = (1/2)^n$. Then, by Problem 2.34 (or directly by an easy argument), find a subsequence x_{k_n} such that

$$|x_{k_{n+1}} - x_{k_n}| \leq p_n \text{ for all } n \in \mathbb{N}.$$

2.37 Let $p,\ q \in \mathbb{N}$ and $p < q$.

1. With the general binomial coefficients as defined in Definition 1.2.13, show that
$$q^k \binom{p}{k} \leq p^k \binom{q}{k}$$
for all $k \in \mathbb{N}$.

2. Show that $(1 + (q/p)x)^p \leq (1 + x)^q$ for all $x \geq 0$.

3. Let $x = 1/n$ and $(q/p)-1 = r > 0$. Transform the last inequality algebraically to obtain
$$\frac{1}{(n+1)^{1+r}} \leq \frac{1}{r}\left(\frac{1}{n^r} - \frac{1}{(n+1)^r}\right)$$
for all $n \in N$.

Solution. If $j \in \mathbb{N}$ then $q(p-j) \leq p(q-j)$. Hence

$$\begin{aligned} k!q^k \begin{pmatrix} p \\ k \end{pmatrix} &= (qp)(q(p-1))(q(p-2))\cdots(q(p-k+1)) \\ &\leq (pq)(p(q-1))(p(q-2))\cdots(p(q-k+1)) \\ &= k!p^k \begin{pmatrix} q \\ k \end{pmatrix}. \end{aligned}$$

This proves the first part. For the second part we use the binomial theorem. Let $k \in \mathbb{Z}^+$. The kth terms for the expansions of $(1 + (q/p)x)^p$ and $(1+x)^q$ are, respectively,

$$(q/p)^k \begin{pmatrix} p \\ k \end{pmatrix} x^k \text{ and } \begin{pmatrix} q \\ k \end{pmatrix} x^k.$$

We see that each term of the expansion for $(1 + (q/p)x)^p$ is dominated by the corresponding term of the expansion for $(1+x)^q$. Hence the second part follows.

For the last part multiply both sides of the inequality

$$\frac{1}{(n+1)^{1+r}} \leq \frac{1}{r}\left(\frac{1}{n^r} - \frac{1}{(n+1)^r}\right)$$

by $r(n+1)^r$. We obtain, with $r = (q/p) - 1$ and $x = 1/n$,

$$\frac{r}{n+1} \leq \left(1 + \frac{1}{n}\right)^r - 1 \text{ or } \frac{r+1+n}{n+1} \leq (1+x)^{q/p}\frac{n}{n+1}.$$

This is easily reduced to the inequality in the second part.

2.5 SERIES OF REAL NUMBERS

2.39 For each $n \geq 1$, let

$$a_n = \frac{n!(3n-1)^2}{1 \cdot 3 \cdot \cdots \cdot (2n+1)}.$$

Determine if $\sum a_n$ converges.

Solution. We have

$$\frac{a_{n+1}}{a_n} = \frac{(n+1)(3n+2)^2}{(3n-1)^2(2n+3)} \to 1/2.$$

Hence, $\sum a_n$ converges.

2.41 Let a_n be a decreasing sequence of positive numbers such that $\sum a_n$ converges. Show that $\lim na_n = 0$.

Solution. Let $\epsilon > 0$. Since $\sum a_n$ converges, there is some N such that for all $m \geq N$, we have

$$a_N + \cdots + a_m < \epsilon.$$

As (a_k) is decreasing, we have $(m - N + 1)a_m \leq a_N + \cdots + a_m$. Hence,

$$ma_m \leq a_N + \cdots + a_m + (N-1)a_m < \epsilon + (N-1)a_m$$

Again, since $\sum a_k$ converges, we have $a_k \to 0$. Hence, for all m sufficiently large, $a_m < \epsilon/N$. Thus, for all $m \geq N$ sufficiently large, we have

$$ma_m \leq \epsilon + (N-1)\epsilon/N \leq 2\epsilon.$$

Since $a_k > 0$, it follows that $\lim_m ma_m = 0$.

2.43 Suppose that $\sum |a_k|$ converges. Show that $\sum |a_k|^p$ converges for all $p > 1$.

Solution. Since $\sum |a_k|$ converges, we have $|a_k| \to 0$. Hence, for all k sufficiently large, $|a_k| < 1$. Thus, since $p > 1$, we have $|a_k|^p \leq |a_k|$ for all k sufficiently large. Hence, $\sum |a_k|^p$ must also converge.

2.45 Suppose that $\sum a_k$ converges. What can we say about the convergence of

$$\sum \frac{1}{1+a_k^2}?$$

Solution. Since $\sum a_k$ converges, we have $a_k \to 0$. This implies that $\frac{1}{1+a_k^2} \to 1 \neq 0$. Hence,

$$\sum \frac{1}{1+a_k^2}$$

diverges.

2.47 Does the series $\sum \sin(1/k)$ converge?

Solution. No. This is because $\sum 1/k$ diverges and

$$\lim_k \frac{\sin(1/k)}{1/k} = 1.$$

2.49 Let r_n be a sequence in $\mathbb{R}$. Assume that there is an $R \in \mathbb{R}$, $0 < R$, such that $|r_n| \leq R^n$ for all $n \in \mathbb{N}$. Show that

$$s_n = \sum\nolimits_{k=0}^{n} \frac{r_k}{k!}$$

is a convergent sequence.

Solution. Find an $N \in \mathbb{N}$ such that $(R/N) = p < 1$. Let $M = (R^N/N!)$. If $k \in \mathbb{N}$ then

$$\begin{aligned} \frac{|r_{N+k}|}{(N+k)!} &\leq \frac{R^{N+k}}{(N+k)!} \\ &= \frac{R^N}{N!} \, \frac{R^k}{(N+1)\cdots(N+k)} \\ &\leq \frac{R^N}{N!} p^k \leq M\, p^{N+k}, \end{aligned}$$

where $M = R^N/(p^N\, N!)$. Then Example 2.3.15 shows that s_n converges.

2.51 Binary expansions. Let $r \in \mathbb{R}$ and $0 \leq r \leq 1$. A sequence b_n is called a *binary expansion* of r if $b_n = 0$ or $b_n = 1$ for each $n \in \mathbb{N}$ and if the sequence

$$s_n = \sum\nolimits_{k=1}^{n} b_k 2^{-k}$$

converges to r. Show that each $r \in [0, 1]$ has a binary expansion. Show that the binary expansion is unique, except for the numbers of the form $r = k2^{-n}$, for some $n, k \in \mathbb{N}$ with $0 \leq k \leq 2^n$. Show that for numbers of this type there are exactly two binary expansions.

Solution. For each $n \in \mathbb{N}$ let $\mathcal{B}_n$ be the set of 2^n intervals $I_{ni} = [i/2^n, (i-1)/2^n)$, $i = 1, \ldots, 2^n$. Then $\mathcal{B}_n$ is a *partition* of $I = [0, 1)$ in the sense that the intervals in $\mathcal{B}_n$ are pairwise disjoint and their union is I. Let $r \in I$ be fixed. Then there is a unique interval $J_n \in \mathcal{B}_n$ that contains r. Denote the end points of J_n by s_n and $t_n = s_n + 2^{-n}$. Hence

$$r \in J_n = [s_n, s_n + 2^{-n}) \text{ and } s_n \leq r < s_n + 2^{-n}.$$

This implies that $\lim_n s_n = r$. We claim that there is a unique sequence b_n such that $b_n = 0$ or $b_n = 1$ and $s_n = \sum_{k=1}^{n} b_k 2^{-k}$ for all $n \in \mathbb{N}$. This follows from an easy induction argument. In fact $s_1 = b_1 2^{-1}$ where $b_1 = 0$ or $b_1 = 1$ and $s_{n+1} = s_n + b_{n+1}2^{-n-1}$ where $b_{n+1} = 0$ or $b_{n+1} = 1$. This shows that each $r \in [0, 1)$ has a binary expansion. We see that $r = 1$ has also a binary expansion such that $b_n = 1$ for all $n \in \mathbb{N}$ and $s_n = 1 - 2^{-n}$. Hence all $r \in [0, 1]$ has a binary expansion. Conversely, we see easily that any $s_n = \sum_{k=1}^{n} b_k 2^{-k}$, $b_k = 0$ or $b_k = 1$, is the binary expansion of a number $r \in [0, 1]$.

Now let $s_n = \sum_{k=1}^{n} b_k 2^{-k}$ be the binary expansion of a number $r \in [0, 1]$. Then we see that $s_n \le r \le s_n + 2^{-n}$. Assume that r has two different binary expansions $s_n = \sum_{k=1}^{n} b_k 2^{-k}$ and $s'_n = \sum_{k=1}^{n} b'_k 2^{-k}$. Let $m \in \mathbb{N}$ be the first integer such that $s_m \ne s'_m$. Without loss of generality assume that $s'_m < s_m$. In this case we see that $s_m = s'_m + 2^{-m}$. Hence $s'_m \le r \le s'_m + 2^{-m} = s_m \le r \le s_m + 2^{-m}$. Therefore $r = s_m$ is the only possibility for r. This is a number of the form $k2^{-n}$, $n \in \mathbb{N}$, $0 \le k < 2^{-n}$. Note that, contrary to the statement in the problem, the numbers $r = 0$ and $r = 1$ have only one binary expansions, as specified in this problem.

2.6 INTERVALS AND CONNECTED SETS

2.53 Let E be a finite subset of $\mathbb{R}$. Show that $\overline{E} = E$.

Solution. Suppose that $E = \{x_1, \ldots, x_m\}$. Let $r = \min\{\, |x_i - x_j| \mid 1 \le i \ne j \le m \,\}$. Let $a \in \overline{E}$. Then there is a sequence (u_n) in E such that $u_n \to a$. Hence, (u_n) is Cauchy. Thus, there is some N such that $|u_p - u_q| < r$ for all $p, q \ge N$. It follows that $u_N = u_{N+1} = u_{N+2} \cdots$. Hence, the sequence (u_n) must converge to u_N. Hence, $a = u_N$. But $u_N \in E$ so that $a \in E$. Thus, $\overline{E} \subset E$.

2.55 What is $\partial\mathbb{Q}$?

Solution. Let $x \in \mathbb{R}$. Then every open interval containing x contains some $r \in \mathbb{Q}$ and some s not in $\mathbb{Q}$. Hence, $r \in \partial\mathbb{Q}$, Thus, $\partial\mathbb{Q} = \mathbb{R}$.

2.57 Show that an intersection of finitely many closed sets in $\mathbb{R}$ is closed.

Solution. Let $A = \cap A_i$, where each A_i is a closed set. We will show that $A = \overline{A}$. Then it will follow from Theorem 2.6.16 that A is closed. Now $A \subset \overline{A}$ is true for any set. Conversely assume that $x \in \overline{A}$. Then Remarks 2.6.8 show that if $r > 0$ then $(x - r,\, x + r)$ intersects $A = \cap_i A_i$. Therefore $(x - r,\, x + r)$ intersects each A_i. Since A_i is closed this means that $x \in A_i$ for each i. Hence $x \in A = \cap_i A_i$. This shows that $\overline{A} \subset A$. Consequently $A = \overline{A}$. Then Corollary 2.6.16 shows that A is closed. Note that the fact that $A = \cap_i A_i$ is a finite intersection is not used in these arguments. Hence the intersection of any family of closed sets is also closed.

2.59 Show that the intersection of any family of intervals is again an interval. Also show that the intersection of any family of closed intervals is again a closed interval. Give examples to show that the intersection of a family of open intervals can be any type of interval (open, closed, or half-open).

Solution. By Definition 2.61 a set $I \subset \mathbb{R}$ is an interval if $a, b \in I$ and if $a < c < b$ then also $c \in I$. Let A be the intersection of a family $\mathcal{I}$ of intervals. Let $a, b \in A$ and $a < c < b$. Let $I \in \mathcal{I}$. Then $a, b \in I$, and therefore also $c \in I$. Therefore c belongs to every interval in the family $\mathcal{I}$. Hence $c \in A$. This shows that A is an interval.

Now assume that each $I \in \mathfrak{I}$ is a (bounded and) closed interval. Assume that their intersection A is not empty. Let $b = \sup A$ be the final point of A. We claim that $b \in A$. Assume that $b \notin A$. In this case there is also a $c \in A$ such that $c < b$. Let $I \in \mathfrak{I}$. Since $A \subset I$ we see that $c \in I$ and $b = \sup A \leq b' = \sup I$. But $b' \in I$, since I is closed. Therefore $b \in I$, since $c < b \leq b'$. Hence b belongs to every $I \in \mathfrak{I}$. This shows that $b \in A$. Similarly we show that $\inf A = a \in A$. Hence A is a closed interval.

To show that the intersection of open intervals can be any type of interval consider $I_n = (0, 1 + (1/n))$, $n \in \mathbb{N}$. Then we see that $\cap_n I_n = (0, 1]$. Examples for other types are similar.

2.61 Let I_n be a sequence of bounded and closed intervals. Assume that the intersection $\cap_{k=1}^{n} I_k$ is nonempty for each $k \in \mathbb{N}$. Show that $\cap_{n \in \mathbb{N}} I_n$ is also nonempty.

Solution. For each $n \in \mathbb{N}$ choose point $x_n \in \cap_{k=1}^{n} I_k$. Then x_n is a bounded sequence since it is contained in I_1. Use the Bolzano-Weierstrass theorem, Theorem 2.4.5, to obtain a convergent subspace of x_n converging to a point x. Then we see that $x \in I_n$ for each $n \in \mathbb{N}$, since I_n contains all the terms of this subspace coming after x_n. Then I_n also contains the limit point x since it is a closed interval. Therefore $x \in \cap_{n \in \mathbb{N}} I_n$ and this intersection is not empty.

2.63 Let I be an interval. Let $A \subset \mathbb{R}$. If I contains points both from A and from its complement $A^c = \mathbb{R} \setminus A$, then show that I also contains points from the boundary of A.

Solution. Let $a \in A \cap I$ and $b \in A^c \cap I$. Without loss of generality assume that $a < b$. Let $E = \{ x \mid a \leq x \leq b \text{ and } x \in A \}$. Then E is a bounded set in $\mathbb{R}$ and $c = \sup E$ exists and $a \leq c \leq b$. Therefore $c \in I$. We claim that $c \in \partial A$.

Distinguish two cases. If $c \in A$ then $c \neq b$ and, therefore $a \leq c < b$. In this case the interval $(c, b]$ cannot contain any points from A, by the definition of c. Hence $(c, b] \subset A^c$. Therefore, if $r > 0$, then $(c - r, c + r)$ intersects A, since $c \in A$, and also it intersects A^c, since it intersects $(c, b]$. Hence $c \in I \cap \partial A$.

If $c \in A^c$ then $a < c$ and $[a, c) \subset A$, by the definition of c. Therefore, if $r > 0$, then $(c - r, c + r)$ intersects A^c, since $c \in A^c$, and also it intersects A, since it intersects $[a, c)$. Hence $c \in I \cap \partial A$.

CHAPTER 3

VECTOR FUNCTIONS

3.1 VECTOR SPACES: THE BASICS

3.1 Let $T : X \to Y$ be a linear mapping. Let U be a subspace of X. Show that if $\dim U = \dim T(U)$, then T is one-to-one on U.

Solution. If T is not one-to-one on U then there is a nonzero $\mathbf{u}_1 \in U$ such that $T\mathbf{u}_1 = \mathbf{0}$. Complete $\{\mathbf{u}_1\}$ to a basis $\{\mathbf{u}_1, \ldots, \mathbf{u}_k\}$ for U. Then we see that $\{T\mathbf{u}_1, \ldots, T\mathbf{u}_k\}$ spans TU and contains less than k nonzero vectors. Hence $\dim TU < k = \dim U$.

3.3 Let X be a vector space. Show that if $\mathbf{x} \in X$ is nonzero and s, t are distinct scalars, then $s\mathbf{x} \neq t\mathbf{x}$.

Solution. If $s\mathbf{x} = t\mathbf{x}$, then $(s - t)\mathbf{x} = \mathbf{0}$. So, if $s - t \neq 0$, then

$$\mathbf{x} = 1\mathbf{x} = \frac{1}{s-t}((s-t)\mathbf{x}) = \frac{1}{s-t}\mathbf{0} = \mathbf{0}.$$

Analysis in Vector Spaces.
By M. A. Akcoglu, P. F. A. Bartha and D. M. Ha

3.5 Let $\mathbf{u} = (1, 2, 3), \mathbf{v} = (2, 5, -4)$. Find all real numbers a such that $(-2, a, 7)$ is a linear combination of $\mathbf{u}$ and $\mathbf{v}$.

Solution. Suppose $(-2, a, 7)$ is a linear combination of $\mathbf{u}$ and $\mathbf{v}$. Then there are scalars s and t such that $(-2, a, 7) = s(1, 2, 3) + t(2, 5, -4)$. Hence,

$$-2 = s + 2t \quad a = 2s + 5t \quad \text{and} \quad 7 = 3s - 4t.$$

The first and the last equation together give $s = 7/5, t = -17/10$. Hence, $a = 2s + 5t = 57/10$.

3.7 Give an example of a nonempty subset A of $\mathbb{R}^2$ such that $A + A \neq 2A$. Find all subsets B of $\mathbb{R}^2$ with $tB + sB \subset B$ for all scalars s and t.

Solution. Take $A = \{(1, 0), (0, 1)\}$. Then $2A = \{(2, 0), (0, 2)\}$ but $A + A = \{(2, 0), (1, 1), (0, 2)\}$.

Now, suppose that B is a subset of $\mathbb{R}^2$ and $tB + sB \subset B$ for all scalars s and t. Either $B = \{\mathbf{0}\}$ or there is some nonzero $\mathbf{u}_0 \in B$. Suppose that the latter holds. Then

$$\{\, \alpha \mathbf{u}_0 \mid \alpha \in \mathbb{R} \,\} \subset B.$$

Let $E = \{\, \alpha \mathbf{u}_0 \mid \alpha \in \mathbb{R} \,\}$. If $E \neq B$, then there is some $\mathbf{v}_0 \in \mathbb{R}^2$ such that $\mathbf{v}_0 \neq \alpha \mathbf{u}_0$ for any scalar α and $\mathbf{v}_0 \in B$. Let $\mathbf{u}_0 = (a, b), \mathbf{v}_0 = (c, d)$. We claim that $bc \neq ad$. For, if $bc = ad$, then $a \neq 0$ and $c \neq 0$ so that $b/a = d/c$ and this would imply that $\mathbf{v}_0 \in E$, a contradiction. Put

$$s = \frac{b}{bc - ad}, \qquad t = \frac{d}{ad - bc}.$$

Then $t\mathbf{u}_0 + s\mathbf{v}_0 = (1, 0)$. Similarly, put

$$t' = \frac{a}{ad - bc}, \qquad s' = \frac{c}{bc - ad}.$$

Then $s'\mathbf{u}_0 + t'\mathbf{v}_0 = (0, 1)$. Thus, if $E \neq B$, then B contains $(1, 0)$ and $(0, 1)$. Since $tB + sB$ for all scalars s and t, it would then follow that $B = \mathbb{R}^2$. Hence, either $B = \{\mathbf{0}\}$, or $B = \{\, \alpha \mathbf{u}_0 \mid \alpha \in \mathbb{R} \,\}$ for some nonzero $\mathbf{u}_0 \in \mathbb{R}^2$, or $B = \mathbb{R}^2$.

3.9 For each $a \in \mathbb{R}$, let $U_a = \{\, (x, y, z) \mid a|x| = x + y + z \,\}$. Find all a for which U_a is a subspace of $\mathbb{R}^3$.

Solution. If $a = 0$, then it is clear that U_a is a subspace of $\mathbb{R}^3$. Now, assume that $a \neq 0$. Let $y, z \in \mathbb{R}$ be such that $a = 1 + y + z$. Then $\mathbf{u} = (1, y, z) \in U_a$ but $-\mathbf{u} \notin U_a$. Hence U_a is not a subspace.

3.11 Let $\mathbf{u}_1, \ldots, \mathbf{u}_n$ be vectors in a vector space X. Let

$$U = \{\, \mathbf{a} = (a_1, \ldots, a_n) \in \mathbb{R}^n \mid a_1\mathbf{u}_1 + \cdots + a_n\mathbf{u}_n = \mathbf{0} \,\},$$

where $\mathbf{0}$ is the zero vector of X. Show that U is a subspace of $\mathbb{R}^n$.

Solution. It is clear that the zero vector of $\mathbb{R}^n$ is in U. Let $\mathbf{a}, \mathbf{b}$ be in U and let s, t be scalars. Then

$$\begin{aligned}(sa_1 + tb_1)\mathbf{u}_1 + \cdots + (sa_n + tb_n)\mathbf{u}_n &= s(a_1\mathbf{u}_1 + \cdots + a_n\mathbf{u}_n) \\ &\quad + t(b_1\mathbf{u}_1 + \cdots + b_n\mathbf{u}_n) \\ &= s\mathbf{0} + t\mathbf{0} = \mathbf{0}.\end{aligned}$$

3.13 Let A be a nonempty subset of a vector space X. Let t be a nonzero scalar. Is it true that Span $(tA) =$ Span A?

Solution. Yes. Let $\mathbf{x} \in$ Span A. Then $\mathbf{x} = c_1\mathbf{u}_1 + \cdots + c_n\mathbf{u}_n$ for some $\mathbf{u}_1, \ldots, \mathbf{u}_n$ in A and scalars $c_1, \ldots, c_n$. Let $\mathbf{v}_1 = t\mathbf{u}_1, \ldots, \mathbf{v}_n = t\mathbf{u}_n$. Then $\mathbf{v}_1, \ldots, \mathbf{v}_n$ are are all in tA, and hence,

$$\mathbf{x} = \frac{c_1}{t}\mathbf{v}_1 + \cdots + \frac{c_n}{t}\mathbf{v}_n \in \text{Span }(tA).$$

Conversely, let $\mathbf{y} \in$ Span (tA). Then there are $\mathbf{w}_1, \ldots, \mathbf{w}_m$ in tA and scalars $d_1, \ldots, d_m$ such that $\mathbf{y} = d_1\mathbf{w}_1 + \cdots + d_m\mathbf{w}_m$. Now, each $\mathbf{w}_i$ is $t\mathbf{u}_i$ for some $\mathbf{u}_i$ in A. Hence,

$$\mathbf{y} = (d_1 t)\mathbf{u}_1 + \cdots + (d_m t)\mathbf{u}_m \in \text{Span } A.$$

3.15 Let A and B be two subsets in a vector space X. Does Span $A \subset$ Span B imply that $A \subset B$?

Solution. No. Choose $\mathbf{v}_0 \neq \mathbf{0}$ and let $A = \{\mathbf{0}, \mathbf{v}_0\}, B = \{\mathbf{v}_0\}$. Then

$$\text{Span } A \subset \text{Span } B, \text{ but } A \not\subset B.$$

3.17 Let A and B be two subsets in a vector space X. Show that

$$\text{Span }(A + B) \subset \text{Span } A + \text{Span } B.$$

Solution. We have $A \subset$ Span A and $B \subset$ Span B. Thus, $A + B \subset$ Span $A +$ Span B. Since Span $A +$ Span B is a subspace and Span $(A + B)$ is a smallest subspace containing $A + B$, we must have Span $(A + B) \subset$ Span $A +$ Span B.

3.19 Let $T : \mathbb{R}^2 \to \mathbb{R}^3$ be defined by

$$T(x, y) = (x - y, y - 3x, x + |y|) \text{ for all } (x, y) \in \mathbb{R}^2.$$

Is T linear?

Solution. No. We have $T(1,0) + T(-1,0) \neq T(0,0)$.

3.21 Let $T \in L(X,Y)$ and let $\mathbf{y}_0 \in Y$ be such that

$$\{\, \mathbf{x} \in X \mid T\mathbf{x} = \mathbf{y}_0 \,\}$$

is a subspace of X. Show that $\mathbf{y}_0 = \mathbf{0}$.

Solution. Assume that $\{\, \mathbf{x} \in X \mid T\mathbf{x} = \mathbf{y}_0 \,\}$ is a subspace of X. Then it must contain $\mathbf{0}$. Hence, $\mathbf{y}_0 = T\mathbf{0}$, and thus, since T is linear, $\mathbf{y}_0 = \mathbf{0}$.

3.23 Let T, S be in $L(X,Y)$. Let $U = \{\, \mathbf{x} \in X \mid T\mathbf{x} = S\mathbf{x} \,\}$. Show that U is a subspace of X.

Solution. Let $L = T - S$. Then $L \in L(X,Y)$. Also,

$$U = \{\, \mathbf{x} \in X \mid L\mathbf{x} = \mathbf{0} \,\} = \text{Ker } L.$$

Hence, U is a subspace of X by Lemma 3.1.29.

3.25 Let $a \in \mathbb{R}$. Consider the linear map $T : \mathbb{R}^3 \to \mathbb{R}^3$ given by

$$T(x,\, y,\, z) = (x,\, ax + y,\, z) \;\text{ for all }\; (x,\, y,\, z) \in \mathbb{R}^3.$$

Show that T is invertible and find its inverse map $T^{-1} : \mathbb{R}^3 \to \mathbb{R}^3$.

Solution. First, T is one-to-one on X since Ker $T = \{\mathbf{0}\}$. Also, for any $(u,\, v,\, w) \in \mathbb{R}^3$, we have $T(u,\, v - au,\, w) = (u,\, v,\, w)$. Thus, the range of T is all of $\mathbb{R}^3$. Hence, T is invertible. In fact, $T^{-1}(u,\, v,\, w) = (u,\, v - au,\, w)$ for all $(u,\, v,\, w) \in \mathbb{R}^3$.

3.27 Let $R : X \to Y$ and $S : Y \to Z$ be isomorphisms. Show that $SR : X \to Z$ is also an isomorphism and $(SR)^{-1} = R^{-1}S^{-1} : Z \to X$.

Solution. This follows directly from the definitions. The following is a more formal argument. Since A, B are invertible, the composition $A^{-1}B^{-1}$ exists. Also,

$$\begin{aligned} (BA)(A^{-1}B^{-1}) &= ((BA)A^{-1})B^{-1} = (B(AA^{-1}))B^{-1} \\ &= (BI_Y)B^{-1} = BB^{-1} = I_Z. \end{aligned}$$

Similarly, $(A^{-1}B^{-1})(BA) = I_X$. Hence, $BA : X \to Z$ is invertible and its inverse is $A^{-1}B^{-1} : Z \to X$. (Here A and B need not be linear. The arguments are valid for any invertible maps.)

3.29 Let X be a vector space. Let $T \in L(X,\, X)$ be non-invertible. Show that $CT = \mathbf{0}$ for a nonzero $C \in L(X,\, X)$.

Solution. Define $\phi : L(X,\, X) \to L(X,\, X)$ by $\phi(R) = RT$ for all $R \in L(X,\, X)$. Then ϕ is linear. Also, since T is not invertible, we have $\phi(R) \neq I_X$ for any

$R \in L(X, X)$. Hence, ϕ is also not one-to-one on $L(X, X)$. This implies that $\phi(C) = \mathbf{0}$ for some nonzero $C \in L(X, X)$.

3.31 Is there a linear map $T : X \to X$ such that T is not one-to-one on X but $T^k : X \to X$ is one-to-one for some $k \geq 2$?

Solution. No. If $T : X \to X$ is not 1-1, then here is a nonzero $\mathbf{u} \in X$ such that $T\mathbf{u} = \mathbf{0}$. Thus, $T^k\mathbf{u} = \mathbf{0}$ for all $k \geq 2$. Hence, since T^k is linear, this implies that T^k is not 1-1 on X.

3.33 Let $T \in L(X, Y)$. Let U be a subspace of X such that $U \cap \operatorname{Ker} T = \{\mathbf{0}\}$. Show that $\dim U = \dim T(U)$.

Solution. If $U \cap \operatorname{Ker} T = \{\mathbf{0}\}$, then we see that T is one-to-one on U. Therefore, by Lemma 3.1.30, $\dim U = \dim T(U)$.

3.35 Let X and Y be two vector spaces. Let W be a subspace of X. If

$$\dim Y \geq (\dim X) - (\dim W),$$

then show that there is a $T \in L(X, Y)$ such that $W = \operatorname{Ker} T$.

First solution. Let U be a complementary subspace to W. Hence $X = U \oplus W$ and $\dim U = \dim X - \dim W$. Now find a subspace V of Y such that $\dim V = \dim U = \dim X - \dim W$. This is possible by the assumption. Then there is an isomorphism $R : U \to V$, since these two spaces are of the same dimension. Let (P, Q) be the pair of complementary projections on X associated with the direct sum decomposition $X = U \oplus W$ as in Theorem 3.1.37. Hence $P : X \to X$ is such that $P(X) = U$ and $P(W) = \{\mathbf{0}\}$. Then let $T = RP : X \to Y$. We see that $T\mathbf{x} = \mathbf{0}$ if and only if $\mathbf{x} \in W$. Hence $W = \ker T$.

Second solution. Let $n = \dim X$, $k = \dim W$, and $\ell = n - k$. Choose a basis $\{\mathbf{w}_1, \ldots, \mathbf{w}_k\}$ for W and complete this to a basis $\{\mathbf{w}_1, \ldots, \mathbf{w}_k, \mathbf{u}_1, \ldots, \mathbf{u}_\ell\}$ for X by the addition of ℓ vectors $\{\mathbf{u}_1, \ldots, \mathbf{u}_\ell\}$. Since $\dim Y \geq \ell$ there is a linearly independent set of ℓ vectors $\{\mathbf{v}_1, \ldots, \mathbf{v}_\ell\}$ in Y. Then there is a linear $T : X \to Y$ uniquely defined as $T\mathbf{w}_i = \mathbf{0}$ for $i = 1, \ldots, k$ and $T\mathbf{u}_j = \mathbf{v}_j$ for $j = 1, \ldots, \ell$. We see that $W = \operatorname{Ker} T$.

3.37 Let U, V be subspaces of a finite-dimensional vector space. Let $W = \{(-\mathbf{x}, \mathbf{x}) \mid \mathbf{x} \in U \cap V\}$.

1. Show that W is a subspace of $U \times V$ and W is isomorphic to $U \cap V$.

2. Define $f : U \times V \to U + V$ by $f(\mathbf{u}, \mathbf{v}) = \mathbf{u} + \mathbf{v}$ for all $(\mathbf{u}, \mathbf{v}) \in U \times V$. Show that f is linear. Hence, deduce that

$$\dim(U + V) = \dim U + \dim V - \dim(U \cap V).$$

Solution. 1. It is clear that $W \subset U \times V$. Also, $(\mathbf{0}, \mathbf{0})$, the zero vector of $U \times V$, is in W since $(\mathbf{0}, \mathbf{0}) = (-\mathbf{0}, \mathbf{0})$. Let $\mathbf{w}_1, \mathbf{w}_2$ be in W and let a, b be scalars. Then $\mathbf{w}_1 = (-\mathbf{x}_1, \mathbf{x}_1), \mathbf{w}_2 = (-\mathbf{x}_2, \mathbf{x}_2)$ for some $\mathbf{x}_1, \mathbf{x}_2$ in $U \cap V$. Now, since $U \cap V$ is also a subspace, $a\mathbf{x}_1 + b\mathbf{x}_2$ is in $U \cap V$. Let $\mathbf{v} = a\mathbf{x}_1 + b\mathbf{x}_2$. Then

$$a\mathbf{w}_1 + b\mathbf{w}_2 = (-a\mathbf{x}_1 - b\mathbf{x}_2, a\mathbf{x}_1 + b\mathbf{x}_2) = (-\mathbf{v}, \mathbf{v}) \in W.$$

Thus, W is a subspace of $U \times V$. To see that $W \sim (U \cap V)$, let $h : W \to U \cap V$ be defined by $h(\mathbf{w}) = \mathbf{x}$ whenever $\mathbf{w} = (-\mathbf{x}, \mathbf{x})$ for some $\mathbf{x} \in U \cap V$. Then h is clearly an isomorphism.

2. Let $(\mathbf{u}_1, \mathbf{v}_1), (\mathbf{u}_2, \mathbf{v}_2)$ be in $U \times V$ and let a, b be scalars. Then

$$\begin{aligned} f(a(\mathbf{u}_1, \mathbf{v}_1) + b(\mathbf{u}_2, \mathbf{v}_2)) &= f(a\mathbf{u}_1 + b\mathbf{u}_2, a\mathbf{v}_1 + b\mathbf{v}_2) \\ &= (a\mathbf{u}_1 + b\mathbf{u}_2) + (a\mathbf{v}_1 + b\mathbf{v}_2) \\ &= a(\mathbf{u}_1 + \mathbf{u}_2) + b(\mathbf{v}_1 + \mathbf{v}_2) \\ &= af(\mathbf{u}_1, \mathbf{v}_1) + bf(\mathbf{u}_2, \mathbf{v}_2). \end{aligned}$$

Thus, $f : U \times V \to U + V$ is linear.

Now, it is clear that $f(U \times V) = U + V$. Also, $f(\mathbf{u}, \mathbf{v}) = \mathbf{0}$ if and only if $\mathbf{u} = -\mathbf{v}$. Thus, an element $(\mathbf{u}, \mathbf{v})$ of $U \times V$ is in Ker f if and only if it is of the form $(-\mathbf{v}, \mathbf{v})$ for some $\mathbf{v} \in U \cap V$. Thus, Ker $f = W$. So, we have

$$\begin{aligned} \dim U + \dim V &= \dim(U \times V) \\ &= \dim(U + V) + \dim \operatorname{Ker} f \\ &= \dim(U + V) + \dim W \\ &= \dim(U + V) + \dim(U \cap V). \end{aligned}$$

3.39 Let X and Z be two vector spaces with $(\dim X) \leq (\dim Z)$. Show that there is a vector space Y with the following property. Given any one-to-one linear mapping $T : X \to Z$, there is an isomorphism $R : (X \times Y) \to Z$ such that the restriction of R to X is T. Here X is identified with the subspace of $X \times Y$ consisting of vectors of the form $(\mathbf{x}, \mathbf{0}) \in X \times Y$, with $\mathbf{x} \in X$.

Solution. Let Y be a subspace complementary to X. Let $T : X \to Z$ be a one-to-one linear mapping and let $U = TX$. Then $\dim U = \dim X$ by Lemma 3.1.30. Let V be a subspace complementary to U. Hence

$$\dim Y = \dim Z - \dim X = \dim Z - \dim U = \dim V.$$

Therefore there is an isomorphism $S : Y \to V$. Then define $R : (X \times Y) \to Z$ by $R(\mathbf{x}, \mathbf{y}) = T\mathbf{x} + S\mathbf{y}$. We see that R has the required properties.

3.41 Let X and Z be two vector spaces with $(\dim X) \leq (\dim Z)$. Show that there is a vector space Y with the following property. Given any linear mapping $S : Z \to X$ that maps Z onto X (that is, $T(Z) = X$), there is an isomorphism $L : Z \to (X \times Y)$ such that $S = PL$, where $P : (X \times Y) \to X$ is the coordinate projection onto X. Recall that $P(\mathbf{x}, \mathbf{y}) = (\mathbf{x}, \mathbf{0})$ for all $(\mathbf{x}, \mathbf{y}) \in X \times Y$. Here X is again identified with a subspace of $X \times Y$.

Solution. Let Y be a subspace complementary to X. Let $S : Z \to X$ be a linear transformation that maps Z onto X. Let $V = \text{Ker } S$ and let U be a subspace complementary to V. Then (U, V) is a coordinate system in Z, in the sense of Definition 3.1.42. Let $M : Z \to U$ and $N : Z \to V$ be the corresponding coordinate projections. Theorem 3.1.32 shows that

$$\dim V = \dim Z - \dim(\text{Range } S) = \dim Z - \dim X = \dim Y.$$

Therefore there is an isomorphism $R : V \to Y$. Define $L : Z \to (X \times Y)$ by $L\mathbf{z} = S\mathbf{z} + RN\mathbf{z}$. Then $L : Z \to Z$ is an isomorphism as it maps Z onto Z. Also, $PL\mathbf{z} = P(S\mathbf{z} + RN\mathbf{z}) = PS\mathbf{z}$ since $RN\mathbf{z} \in Y$ and $PY = \{\mathbf{0}\}$.

3.43 Let (U, V_i), $i = 1, 2$, be two coordinate systems in X, in the sense of Definition 3.1.42. Consider the following proposition: a set $\Gamma \subset X$ is a graph in one system if and only if it is a graph in the other system. Show either that this proposition is true or that it is false.

Solution. This proposition is false. As a counterexample let X be the standard xy-plane. Let U be the x-axis, let V_1 be the y-axis, and let V_2 be the line $y = x$. Then we see that V_2 is a graph in the coordinate system (U, V_1), but not a graph in the coordinate system (U, V_2).

3.2 BILINEAR FUNCTIONS

3.45 The *cross product* of two vectors is defined as

$$(x_1, y_1, z_1) \times (x_2, y_2, z_2) = (y_1 z_2 - y_2 z_1, z_1 x_2 - x_1 z_2, x_1 y_2 - y_1 x_2).$$

Show that the cross product, as a function $\mathbb{R}^3 \times \mathbb{R}^3 \to \mathbb{R}^3$, is a bilinear operation. Hence, the cross product is also a product in the sense defined here.

Solution. We verify that $(\mathbf{u} + \mathbf{y}) \times \mathbf{v} = \mathbf{u} \times \mathbf{v} + \mathbf{y} \times \mathbf{v}$ and $(c\mathbf{u}) \times \mathbf{v} = c(\mathbf{u} \times \mathbf{v})$. Hence, the cross product $T : \mathbb{R}^3 \times \mathbb{R}^3 \to \mathbb{R}^3$ is linear in the first variable. We also verify that $(\mathbf{u} \times \mathbf{v}) = -(\mathbf{v} \times \mathbf{u})$ so that T is also linear in the second variable. Hence, T is bilinear.

3.47 Show that $f(x_1, \ldots, x_n) = \sum_{i=1}^{n} \sum_{j=1}^{i} x_i x_j \mathbf{a}_{ij}$ is a general second-degree homogeneous polynomial $f : \mathbb{R}^n \to Y$. Here $\mathbf{a}_{ij} \in Y$ are arbitrary.

Solution. A general second degree homogeneous polynomial $f : \mathbb{R}^n \to Y$ is of the form $f(\mathbf{x}) = B(\mathbf{x}, \mathbf{x})$, where $B : \mathbb{R}^n \times \mathbb{R}^n \to Y$ is a bilinear map. By the previous problem, any such B is of the form

$$B((u_1, \ldots, u_n), (v_1, \ldots, v_n)) = \sum_{i=1}^{n} \sum_{j=1}^{n} u_i v_j \mathbf{a}_{ij}$$

with arbitrary $\mathbf{a}_{ij} \in Y$. Hence,

$$f(\mathbf{x}) = B(\mathbf{x}, \mathbf{x}) = \sum_{i=1}^{n} \sum_{j=1}^{n} x_i x_j \mathbf{a}_{ij}$$

with arbitrary $\mathbf{a}_{ij} \in Y$.

3.49 Let X be a vector space with basis $\{\mathbf{x}_1, \ldots, \mathbf{x}_k\}$. Then every bilinear map $S : X \times \mathbb{R}^n \to \mathbb{R}^m$ is of the form

$$S(\mathbf{x}, \mathbf{y}) = c_{x1} A_1 \mathbf{y} + \cdots + c_{xk} A_k \mathbf{y} \quad \text{for all } \mathbf{y} \in \mathbb{R}^n,$$

where $A_1, \ldots, A_k$ are $m \times n$ matrices and $c_{x1}, \ldots, c_{xk}$ are scalars such that

$$\mathbf{x} = c_{x1} \mathbf{x}_1 + \cdots + c_{xk} \mathbf{x}_k.$$

Solution. For each $j = 1, \ldots, k$, the function $\mathbf{y} \mapsto T(\mathbf{x}_j, \mathbf{y})$ is a linear map from $\mathbb{R}^n \to \mathbb{R}^m$. Let A_j be its standard matrix. Then whenever $\mathbf{x} = c_{x_1} \mathbf{x}_1 + \cdots + c_{xk} \mathbf{x}_k$, we have

$$\begin{aligned} T(\mathbf{x}, \mathbf{y}) &= c_{x1} T(\mathbf{x}_1, \mathbf{y}) + \cdots + c_{xk} T(\mathbf{x}_k, \mathbf{y}) \\ &= c_{x1} A_1 \mathbf{y} + \cdots + c_{xk} A_k \mathbf{y} \quad \text{for all } \mathbf{y} \in \mathbb{R}^n. \end{aligned}$$

3.51 Let $T : X \times Y \to Z$ be bilinear. Suppose that

$$(\dim Y)(\dim Z) < \dim X.$$

Show that there is a nonzero $\mathbf{x}_0 \in X$ such that

$$T(\mathbf{x}_0, \mathbf{y}) = \mathbf{0} \quad \text{for all } \mathbf{y} \in Y.$$

Solution. Let $B = \{\mathbf{x}_1, \ldots, \mathbf{x}_n\}$ be a basis for X and define $T_k : Y \to Z$ by $T_k(\mathbf{y}) = T(\mathbf{x}_k, \mathbf{y})$ for all $\mathbf{y} \in Y$ and all $k = 1, \ldots, n$. Then $T_1, \ldots, T_n$ are linear maps from Y into Z. Hence, $U = \{T_1, \ldots, T_n\} \subset L(Y, Z)$. Since $\dim L(Y, Z) = (\dim Y)(\dim Z) < \dim X = n$, we see that U is a linearly dependent subset of

$L(Y, Z)$. Thus, there are scalars $c_1, \ldots, c_n$, not all 0, such that $c_1T_1 + \cdots + c_nT_n = \mathbf{0}$. Thus,

$$\mathbf{0} = c_1T_1(\mathbf{y}) + \cdots + c_nT_n(\mathbf{y}) = T(c_1\mathbf{x}_1 + \cdots + c_n\mathbf{x}_n, \mathbf{y}) \quad \text{for all } \mathbf{y} \in Y.$$

Put $\mathbf{x}_0 = c_1\mathbf{x}_1 + \cdots + c_n\mathbf{x}_n$. Since B is a basis for X and $c_1, \ldots, c_n$ are not all 0, we see that $\mathbf{x}_0 \neq \mathbf{0}$.

3.3 MULTILINEAR FUNCTIONS

3.53 Define $\varphi : \mathbb{R}^3 \times \mathbb{R}^3 \times \mathbb{R}^3 \to \mathbb{R}$ as $\varphi(\mathbf{x}, \mathbf{y}, \mathbf{z}) = \langle \mathbf{x} \times \mathbf{y}, \mathbf{z} \rangle$, where $\mathbf{x} \times \mathbf{y}$ is the usual cross-product of $\mathbf{x}$ and $\mathbf{y}$, and $\langle , \rangle$ is the usual inner product operation. Show that φ is a multilinear function. Find a linear function $T : \mathbb{R}^3 \to BL(\mathbb{R}^3 \times \mathbb{R}^3, \mathbb{R})$ such that $\varphi(\mathbf{x}, \mathbf{y}, \mathbf{z}) = (T\mathbf{x})(\mathbf{y}, \mathbf{z})$ for all $(\mathbf{x}, \mathbf{y}, \mathbf{z}) \in \mathbb{R}^3 \times \mathbb{R}^3 \times \mathbb{R}^3$.

Solution. Both the cross product and the inner product are bilinear operations. Then an easy check shows that the mixed product $\varphi(\mathbf{x}, \mathbf{y}, \mathbf{z}) = \langle \mathbf{x} \times \mathbf{y}, \mathbf{z} \rangle$ is a multilinear function. For $\mathbf{x} \in \mathbb{R}^3$ define $T\mathbf{x} : (\mathbb{R}^3 \times \mathbb{R}^3) \to \mathbb{R}$ by $(T\mathbf{x})(\mathbf{y}, \mathbf{z}) = \langle \mathbf{x} \times \mathbf{y}, \mathbf{z} \rangle$. Then $T : \mathbb{R}^3 \to BL(\mathbb{R}^3 \times \mathbb{R}^3, \mathbb{R})$ is a linear function. This follows from the multilinearity of the mixed product.

3.55 Define $\varphi : \mathbb{R}^3 \times \mathbb{R}^3 \times \mathbb{R}^3 \to \mathbb{R}^3$ as $\varphi(\mathbf{x}, \mathbf{y}, \mathbf{z}) = (\mathbf{x} \times \mathbf{y}) \times \mathbf{z}$. Show that φ is a multilinear function. Find a linear function $T : \mathbb{R}^3 \to BL(\mathbb{R}^3 \times \mathbb{R}^3, \mathbb{R}^3)$ such that $\varphi(\mathbf{x}, \mathbf{y}, \mathbf{z}) = (T\mathbf{x})(\mathbf{y}, \mathbf{z})$ for all $(\mathbf{x}, \mathbf{y}, \mathbf{z}) \in \mathbb{R}^3 \times \mathbb{R}^3 \times \mathbb{R}^3$.

3.57 Let $X_1, \ldots, X_k$ be vector spaces and let $T : X_1 \times \cdots \times X_k \to X_1 \times \cdots \times X_k$ be multilinear. Suppose that there are linear maps $T_j : X_j \to X_j$ such that $T(\mathbf{x}_1, \ldots, \mathbf{x}_k) = (T_1\mathbf{x}_1, \ldots, T_k\mathbf{x}_k)$. If $k \geq 2$, must $T = \mathbf{0}$?

Solution. Yes. Fix any j with $1 \leq j \leq k$. Let $\mathbf{x} \in X_j$ be arbitrary. Let $\mathbf{u} = (\mathbf{0}, \ldots, \mathbf{x}, \ldots, \mathbf{0}) \in X_1 \times \cdots \times X_k$. Then we see easily, directly or from the previous problem, $T\mathbf{u} = \mathbf{0}$, the zero vector in $X_1 \times \cdots \times X_k$. Thus,

$$(\mathbf{0}, \ldots, \mathbf{0}) = \mathbf{0} = T(\mathbf{u}) = (T_1\mathbf{0}, \ldots, T_j\mathbf{x}, \ldots, T_k\mathbf{0}).$$

3.59 Let $T : \mathbb{R}^3 \times \mathbb{R}^2 \times \mathbb{R}^5 \to \mathbb{R}^4$ be multilinear. Show that there is a multilinear map $\varphi : \mathbb{R}^3 \times \mathbb{R}^2 \to M_{4\times 5}$ such that

$$T(\mathbf{x}, \mathbf{y}, \mathbf{z}) = \varphi(\mathbf{x}, \mathbf{y})\mathbf{z} \quad \text{for all } (\mathbf{x}, \mathbf{y}, \mathbf{z}) \in \mathbb{R}^3 \times \mathbb{R}^2 \times \mathbb{R}^5.$$

Here, we identify each vector in $\mathbb{R}^m$ with an $m \times 1$ column matrix. Thus, the right-hand side of the equation above is interpreted as the product of a 4×5 matrix and a 5×1 column matrix. The product obtained is a 4×1 matrix, which is identified with the corresponding vector in $\mathbb{R}^4$.

Solution. Let $T : \mathbb{R}^3 \times \mathbb{R}^2 \times \mathbb{R}^5 \to \mathbb{R}^4$ be multilinear. For each $(\mathbf{x}, \mathbf{y}) \in \mathbb{R}^3 \times \mathbb{R}^2$, define $T_{(\mathbf{x},\mathbf{y})} : \mathbb{R}^5 \to \mathbb{R}^4$ by

$$T_{(\mathbf{x},\mathbf{y})}(\mathbf{z}) = T(\mathbf{x}, \mathbf{y}, \mathbf{z}) \quad \text{for all } \mathbf{z} \in \mathbb{R}^5.$$

Then $T_{(\mathbf{x},\mathbf{y})}$ is a linear map, and thus, its standard matrix is a 4×5 matrix $A_{(\mathbf{x},\mathbf{y})}$. Thus,

$$T_{(\mathbf{x},\mathbf{y})}(\mathbf{z}) = A_{(\mathbf{x},\mathbf{y})}\mathbf{z}.$$

Since T is multilinear,

$$\begin{aligned} T(a_1\mathbf{x}_1 + a_2\mathbf{x}_2, \mathbf{y}, \mathbf{z}) &= a_1T(\mathbf{x}_1, \mathbf{y}, \mathbf{z}) + a_2T(\mathbf{x}_2, \mathbf{y}, \mathbf{z}) \\ &= a_1A_{(\mathbf{x}_1,\mathbf{y})}\mathbf{z} + a_2A_{(\mathbf{x}_2,\mathbf{y})}\mathbf{z} \\ &= [a_1A_{(\mathbf{x}_1,\mathbf{y})} + a_2A_{(\mathbf{x}_2,\mathbf{y})}]\mathbf{z} \end{aligned}$$

Hence, $A_{(a_1\mathbf{x}_1+a_2\mathbf{x}_2,\mathbf{y})}\mathbf{z} = [a_1A_{(\mathbf{x}_1,\mathbf{y})} + a_2A_{(\mathbf{x}_2,\mathbf{y})}]\mathbf{z}$ so that

$$[a_1A_{(\mathbf{x}_1,\mathbf{y})} + a_2A_{(\mathbf{x}_2,\mathbf{y})}] = A_{(a_1\mathbf{x}_1+a_2\mathbf{x}_2,\mathbf{y})}.$$

Similarly,

$$A_{(\mathbf{x},b_1\mathbf{y}_1+b_2\mathbf{y}_2)} = b_1A_{(\mathbf{x},\mathbf{y}_1)} + b_2A_{(\mathbf{x},\mathbf{y}_2)}.$$

Define $\varphi : \mathbb{R}^3 \times \mathbb{R}^2 \to M_{4\times 5}$ by

$$\varphi(\mathbf{x}, \mathbf{y}) = A_{(\mathbf{x},\mathbf{y})} \quad \text{for all } (\mathbf{x}, \mathbf{y}) \in \mathbb{R}^3 \times \mathbb{R}^2.$$

Then, $T(\mathbf{x}, \mathbf{y}, \mathbf{z}) = \varphi(\mathbf{x}, \mathbf{y})\mathbf{z}$ and it follows right from the preceding equations that φ is multilinear.

3.61 Let $T \in ML(X_1 \times \cdots \times X_k, Z)$ and let $B_1, \ldots, B_k$ be nonempty bases for $X_1, \ldots, X_k$, respectively. Show that if $T(\mathbf{b}_1, \ldots, \mathbf{b}_k) = \mathbf{0}$ for all $(\mathbf{b}_1, \ldots, \mathbf{b}_k) \in B_1 \times \cdots \times B_k$, then $T = \mathbf{0}$.

Solution. For $k = 1$, the assertion is clear. Assume that the assertion holds for some $k \geq 1$. Let $S \in ML(X_1 \times \cdots \times X_k \times X_{k+1}, Z)$ and let $B_1, \ldots, B_k, B_{k+1}$ be nonempty bases for $X_1, \ldots, X_k, X_{k+1}$, respectively. Suppose that

$$S(\mathbf{b}_1, \ldots, \mathbf{b}_k, \mathbf{b}_{k+1}) = \mathbf{0} \text{ for all } (\mathbf{b}_1, \ldots, \mathbf{b}_k, \mathbf{b}_{k+1}) \in B_1 \times \cdots \times B_k \times B_{k+1}$$

and also that $B_{k+1} = \{\mathbf{u}_1, \ldots, \mathbf{u}_m\}$. For each $j = 1, \ldots, m$, define

$$\begin{aligned} T_j &: X_1 \times \cdots \times X_k \to Z \text{ by} \\ T_j(\mathbf{x}) &= S(\mathbf{x}_1, \ldots, \mathbf{x}_k, \mathbf{u}_j) \quad \text{for all } \mathbf{x} = (\mathbf{x}_1, \ldots, \mathbf{x}_k) \in X_1 \times \cdots \times X_k. \end{aligned}$$

Then $T_j \in ML(X_1 \times \cdots \times X_k, Z)$ and $T_j(\mathbf{b}_1, \ldots, \mathbf{b}_k) = \mathbf{0}$ for all $(\mathbf{b}_1, \ldots, \mathbf{b}_k) \in B_1 \times \cdots \times B_k$ and all $j = 1, \ldots, m$. Thus, by he induction hypothesis,

$$T_j = \mathbf{0} \quad \text{for all } j = 1, \ldots, m.$$

We will show that $S = \mathbf{0}$. Let $(\mathbf{x}_1, \ldots, \mathbf{x}_k, \mathbf{u}) \in X_1 \times \cdots \times X_k \times X_{k+1}$ be arbitrary. Then $\mathbf{u} \in X_{k+1}$ so that $\mathbf{u} = c_1\mathbf{u}_1 + \cdots + c_m\mathbf{u}_m$ for some scalars $c_1, \ldots, c_m$. Thus, because S is multilinear,

$$\begin{aligned} S(\mathbf{x}_1, \ldots, \mathbf{x}_k, \mathbf{u}) &= \sum_{j=1}^{m} c_j S(\mathbf{x}_1, \ldots, \mathbf{x}_k, \mathbf{u}_j) \\ &= \sum_{j=1}^{m} c_j T_j(\mathbf{x}_1, \ldots, \mathbf{x}_k) = \mathbf{0}. \end{aligned}$$

3.63 Let $X_1, \ldots, X_k, Z$ be vector spaces and suppose that $B_1, \ldots, B_k$ are bases for $X_1, \ldots, X_k$, respectively. Show that

1. Given $\mathbf{b} \in B_1 \times \cdots \times B_k$ and $\mathbf{z} \in Z$, there is a unique multilinear map $T_{\mathbf{b},\mathbf{z}} : X_1 \times \cdots \times X_k \to Z$ such that

$$T_{\mathbf{b},\mathbf{z}}(\mathbf{x}) = \begin{cases} \mathbf{0} & \text{if } \mathbf{x} \in B_1 \times \cdots \times B_k, \mathbf{x} \neq \mathbf{b} \\ \mathbf{z} & \text{if } \mathbf{x} = \mathbf{b}. \end{cases}$$

2. If B is a basis for Z, then a basis for $ML(X_1 \times \cdots \times X_k, Z)$ is the set

$$\{ T_{\mathbf{b},\mathbf{z}} \mid \mathbf{b} \in B_1 \times \cdots \times B_k, \mathbf{z} \in B \},$$

where $T_{\mathbf{b},\mathbf{z}}$ is the unique multilinear map in part 1. It follows that

$$\dim ML(X_1 \times \cdots \times X_k, Z) = (\dim X_1) \cdots (\dim X_k)(\dim Z).$$

3. If $k \geq 2$, then

$$ML(X_1 \times \cdots \times X_k, Z) \sim L(X_1, ML(X_2 \times \cdots \times X_k, Z)).$$

Solution. This problem is an elaboration of the arguments given in Definition 3.3.7. The notations are somewhat different. We will proceed by induction on $k \in \mathbb{N}$. The linear case corresponds to $k = 1$. The main argument for the solution is given in Remarks 3.1.4 concerning the linear case. Let X and Y be two linear spaces. Let C be a basis for X and let E be a basis for Y. Then for each pair $(\mathbf{c}^0, \mathbf{e}^0) \in C \times E$ there is a unique $P(\mathbf{c}^0, \mathbf{e}^0) \in L(X, Y)$ such that $P(\mathbf{c}^0, \mathbf{e}^0)(\mathbf{c}^0) = \mathbf{e}^0$ and $P(\mathbf{c}^0, \mathbf{e}^0)(\mathbf{c}) = \mathbf{0}$ if $\mathbf{c} \in C$ and $\mathbf{c} \neq \mathbf{c}^0$. Also, the set of these transformations

$$P(\mathbf{c}, \mathbf{e}), \text{ where } (\mathbf{c}, \mathbf{e}) \in C \times E,$$

is a basis for $L(X, Y)$. Hence $\dim L(X, Y) = (\dim X) \cdot (\dim Y)$. The induction step will be the application of this argument to the liner space

$$Y = ML(X^k, Z), \text{ where } X^k = X_1 \times \cdots \times X_k.$$

Let B_i be a basis for X_i and U as a basis for Z. Let $B^k = B_1 \times \cdots \times B_k$. As an induction hypothesis assume that $Y = ML(X^k, Z)$ has a basis consisting of $T_{\mathbf{b}^0, \mathbf{u}^0}$, where $(\mathbf{b}^0, \mathbf{u}^0) \in B^k \times U$ and $T_{\mathbf{b}^0, \mathbf{u}^0} \in ML(X^k, Z)$ is defined as

$$T_{\mathbf{b}^0, \mathbf{u}^0}(\mathbf{b}) = \begin{cases} \mathbf{0} & \text{if } \mathbf{b} \in B^k \text{ and } \mathbf{b} \neq \mathbf{b}^0 \\ \mathbf{u}^0 & \text{if } \mathbf{b} = \mathbf{b}^0. \end{cases}$$

Hence $L(X, Y)$ has a basis consisting of $P(\mathbf{c}, T_{\mathbf{b}, \mathbf{u}}) \in L(X, Y)$ defined as

$$P(\mathbf{c}^0, T_{\mathbf{b}^0, \mathbf{u}^0})(\mathbf{c}) = \begin{cases} 0 & \text{if } \mathbf{c} \in C \text{ and } \mathbf{c} \neq \mathbf{c}^0 \\ T_{\mathbf{b}^0, \mathbf{u}^0} & \text{if } \mathbf{c} = \mathbf{c}^0. \end{cases}$$

Therefore, by the definition of $T_{\mathbf{b}^0, \mathbf{c}^0}$ above,

$$P(\mathbf{c}^0, T_{\mathbf{b}^0, \mathbf{u}^0})(\mathbf{c})(\mathbf{b}) = \begin{cases} 0 & \text{if } (\mathbf{b}, \mathbf{c}) \in B^k \times C \text{ and } (\mathbf{b}, \mathbf{c}) \neq (\mathbf{b}^0, \mathbf{c}^0) \\ \mathbf{u}^0 & \text{if } (\mathbf{b}, \mathbf{c}) = (\mathbf{b}^0, \mathbf{c}^0). \end{cases}$$

Now, to conform with the notations of the problem, let $X = X_{k+1}$ and $C = B_{k+1}$. Denote $B^k \times B_{k+1}$ as B^{k+1}. Then we see that $P(\mathbf{c}, T_{\mathbf{b}, \mathbf{u}})$ in $L(X_{k+1}, ML(X^k, Z))$ is the same as $T_{(\mathbf{b}, \mathbf{c}), \mathbf{u}}$ in $ML(X^{k+1}, Z)$. This shows the existence of these multilinear functions and that they form a basis for $ML(X^{k+1}, Z)$.

3.65 Find all homogeneous polynomials of degree three from $\mathbb{R}^2$ to $\mathbb{R}$.

Solution. We would like to find the general form of a symmetric multilinear function $\varphi : \mathbb{R}^2 \times \mathbb{R}^2 \times \mathbb{R}^2 \to \mathbb{R}$. Let $(\mathbf{u}, \mathbf{v})$ be a basis for $\mathbb{R}^2$. Let

$$\varphi(\mathbf{u}, \mathbf{u}, \mathbf{u}) = A, \ \varphi(\mathbf{u}, \mathbf{u}, \mathbf{v}) = B, \ \varphi(\mathbf{u}, \mathbf{v}, \mathbf{v}) = C, \ \varphi(\mathbf{v}, \mathbf{v}, \mathbf{v}) = D.$$

Let $\mathbf{z}_i = (x_i\mathbf{u} + y_i\mathbf{v})$ be three vectors in $\mathbb{R}^2$. Then, by the multilinearity and symmetry of φ,

$$\begin{aligned} \varphi(\mathbf{z}_1, \mathbf{z}_2, \mathbf{z}_3) &= Ax_1x_2x_3 + B(x_1x_2y_3 + x_1y_2x_3 + y_1x_2x_3) \\ &\quad + C(x_1y_2y_3 + y_1x_2y_3 + y_1y_2x_3) + Dy_1y_2y_3. \end{aligned}$$

This is the general form of a homogeneous polynomial of degree three from $\mathbb{R}^2$ to $\mathbb{R}$.

3.4 INNER PRODUCTS

3.67 Suppose that X is any finite dimensional vector space. Show that given a homogeneous polynomial $f : X \to \mathbb{R}$ of degree 2 and any inner product $\langle\ ,\ \rangle$ on X, there is a linear map $L : X \to X$ such that

$$f(\mathbf{x}) = \langle \mathbf{x}, L\mathbf{x} \rangle \quad \text{for all } \mathbf{x} \in X.$$

Solution. Suppose that $f(\mathbf{x}) = Q(\mathbf{x}, \mathbf{x})$ for all $\mathbf{x} \in X$, where $Q : X \times X \to \mathbb{R}$ is multilinear. For each $\mathbf{u} \in X$, the function $T_{\mathbf{u}} : X \to \mathbb{R}$ defined by $T_{\mathbf{u}}(\mathbf{x}) = Q(\mathbf{u}, \mathbf{x})$ is a linear functional. Let $\langle \ , \ \rangle$ be any inner product on X. Then there exists a unique $\mathbf{w_u} \in X$ such that

$$T_u(\mathbf{x}) = \langle \mathbf{x}, \mathbf{w_u} \rangle \quad \text{for all } \mathbf{x} \in X.$$

Define $L : X \to X$ by $L\mathbf{u} = \mathbf{w_u}$. Then L is linear. Also, $Q(\mathbf{u}, \mathbf{x}) = \langle \mathbf{x}, L\mathbf{u} \rangle$. Thus, $f(\mathbf{x}) = Q(\mathbf{x}, \mathbf{x}) = \langle \mathbf{x}, L\mathbf{x} \rangle$ for all $\mathbf{x} \in X$.

3.69 Let X be an inner product space. Let $\mathbf{x}$, $\mathbf{y} \in X$ and $\mathbf{x} \neq \mathbf{0}$. Show that $\|\mathbf{x} + \mathbf{y}\| = \|\mathbf{x}\| + \|\mathbf{y}\|$ if and only if there is an $\alpha \geq 0$ such that $\mathbf{y} = \alpha\mathbf{x}$.

Solution. Let $\mathbf{x}, \mathbf{y}$ be in X. If $\mathbf{y} = \alpha\mathbf{x}$ for some $\alpha \geq 0$, then

$$\|\mathbf{x} + \mathbf{y}\| = \|\mathbf{x} + \alpha\mathbf{x}\| = (1 + \alpha)\,\|\mathbf{x}\| = (1 + |\alpha|)\,\|\mathbf{x}\| = \|\mathbf{x}\| + \|\mathbf{y}\|.$$

Conversely, assume that $\|\mathbf{x} + \mathbf{y}\| = \|\mathbf{x}\| + \|\mathbf{y}\|$ and $\mathbf{x} \neq \mathbf{0}$. Then from the proof of Theorem 3.4.7, we must necessary have equality in the Cauchy Schwartz Inequality. That is,

$$\langle \mathbf{x}, \, \mathbf{y} \rangle = \|\mathbf{x}\|\,\|\mathbf{y}\|. \tag{3.1}$$

An examination of the proof of Theorem 3.4.6 shows that if (3.1) holds, then $\langle t\mathbf{x} + \mathbf{y}, \, t\mathbf{x} + \mathbf{y} \rangle = 0$ for some $t \in \mathbb{R}$. Hence, if (3.1) holds, then $\mathbf{y} = \alpha\mathbf{x}$ for some $\alpha \in \mathbb{R}$, and then (3.1) implies that

$$\alpha\,\|\mathbf{x}\|^2 = |\alpha|\,\|\mathbf{x}\|^2.$$

Hence, since $\|\mathbf{x}\| \neq 0$, we have $\alpha \geq 0$.

3.71 Suppose that $\mathbf{u}, \mathbf{v}$ are vectors in an inner product space X and $\|\mathbf{u}\| = \|\mathbf{v}\|$. Show that $\mathbf{u} - \mathbf{v} \perp \mathbf{u} + \mathbf{v}$.

Solution. Since $\|\mathbf{u}\| = \|\mathbf{v}\|$, we have

$$\begin{aligned} \langle \mathbf{u} - \mathbf{v}, \, \mathbf{u} + \mathbf{v} \rangle &= \|\mathbf{u}\|^2 + \langle \mathbf{u}, \, \mathbf{v} \rangle - \langle \mathbf{v}, \, \mathbf{u} \rangle - \|\mathbf{v}\|^2 \\ &= \|\mathbf{u}\|^2 - \|\mathbf{v}\|^2 = 0. \end{aligned}$$

3.73 Let X be a Euclidean space. Let $f \in L(X, \mathbb{R})$. Then, by Theorem 3.4.21, there is a unique $\mathbf{a}_f \in W$ such that $f(\mathbf{x}) = \langle \mathbf{a}_f, \, \mathbf{x} \rangle$ for all $\mathbf{x} \in X$. Define $T : L(X, \mathbb{R}) \to X$ by $Tf = \mathbf{a}_f$. Show that T is an isomorphism.

Solution. To show that $T : L(X,\mathbb{R}) \to X$ is an isomorphism we show that it is a linear, one-to-one, and onto map. Let $f,\, g \in L(W,\mathbb{R})$ and $a,\, b \in \mathbb{R}$. Then

$$\begin{aligned}\langle aTf + bTg,\, \mathbf{x}\rangle &= a\langle Tf,\, \mathbf{x}\rangle + b\langle Tg,\, \mathbf{x}\rangle \\ &= af(\mathbf{x}) + bg(\mathbf{x}) \\ &= (af + bg)(\mathbf{x}) \\ &= \langle T(af + bg),\, \mathbf{x}\rangle.\end{aligned}$$

Hence $aTf + bTg = T(af + bg)$ and T is linear. To see that T is one-to-one we check that Ker $T = \{\mathbf{0}_{L(X,Y)}\}$. Let $Tf = \mathbf{0}_X$. Then $f(\mathbf{x}) = \langle \mathbf{0}_X,\, \mathbf{x}\rangle = 0$ for all $\mathbf{x} \in X$. Hence $f = \mathbf{0}_{L(X,Y)}$. Finally, to see that $T : L(X,\mathbb{R}) \to X$ is onto, let $\mathbf{a} \in X$. Define $f \in L(X,\, \mathbb{R})$ by $f(\mathbf{x}) = \langle \mathbf{a},\, \mathbf{x}\rangle$. Then we see that $Tf = \mathbf{a}$. Hence $T : L(X,\mathbb{R}) \to X$ is an isomorphism.

3.75 Let X be a Euclidean space. Let $f_1, \ldots, f_m$ be in $L(X,\, \mathbb{R})$, If $m > \dim X$, then show that some f_k is a linear combination of the remaining f_js.

Solution. Problem 3.73 shows that there is an isomorphism between X and $L(X,\, \mathbb{R})$. Hence $\dim L(X,\, \mathbb{R}) = \dim X$. Therefore, if $m > \dim X$, then $f_1, \ldots, f_m$ cannot be linearly independent; one f_k must be a linear combination of the remaining f_j s.

3.77 Let X be an inner product space, and let $\mathbf{a} \in X$ be nonzero. Show that for any scalar c, there is an $f \in L(X,\, \mathbb{R})$ and a $\mathbf{u} \in X$ such that

$$\{\, \mathbf{x} \in X \mid \langle \mathbf{x}, \mathbf{a}\rangle = c \,\} = \text{Ker } f + \mathbf{u}.$$

Solution. Define $f \in L(X,\, \mathbb{R})$ by $f(\mathbf{x}) = \langle \mathbf{a},\, \mathbf{x}\rangle$, $\mathbf{x} \in X$. Since $\mathbf{a} \neq \mathbf{0}$ there is a $\mathbf{u} \in X$ such that $\langle \mathbf{a},\, \mathbf{u}\rangle = c$. Take, for example, $\mathbf{u} = (c/\|\mathbf{a}\|^2)\mathbf{a}$. Then

$$\langle \mathbf{a},\, \mathbf{x}\rangle = c = \langle \mathbf{a},\, \mathbf{u}\rangle$$

if and only if $\langle \mathbf{a},\, \mathbf{x} - \mathbf{u}\rangle = 0$, that is, if and only if $(\mathbf{x} - \mathbf{u}) \in \text{Ker } f$.

3.5 ORTHOGONAL PROJECTIONS

3.79 Let $(W, \langle\;,\;\rangle)$ be a Euclidean space, and let $\mathbf{a}$ be a nonzero vector in W. Let $c \in \mathbb{R}$ and let $E = \{\, \mathbf{w} \in W \mid \langle \mathbf{w}, \mathbf{a}\rangle = c \,\}$. Show that

$$\min_{\mathbf{e}\in E} \|\mathbf{w}_0 - \mathbf{e}\| = \frac{|\langle \mathbf{w}_0, \mathbf{a}\rangle - c|}{\|\mathbf{a}\|} \quad \text{for all } \mathbf{w}_0 \in W.$$

Solution. Set $U = \{\, \mathbf{w} \in W \mid \langle \mathbf{w}, \mathbf{a}\rangle = 0 \,\}$. Then $U^{\perp} = \text{Span}\,\{\mathbf{a}\}$ and $\{\mathbf{a}/\|\mathbf{a}\|\}$ is an orthonormal basis for $U^{\perp}$. Thus, for all $\mathbf{x} \in W$, we have $P_{U^{\perp}}\mathbf{x} = \langle \mathbf{x}, \mathbf{a}\rangle\, \mathbf{a}/\|\mathbf{a}\|^2$.

Therefore

$$\|\mathbf{x} - P_U\mathbf{x}\| = \|P_{U^\perp}\mathbf{x}\| = \frac{|\langle \mathbf{x}, \mathbf{a}\rangle|}{\|\mathbf{a}\|}. \tag{3.2}$$

Now, $E \neq \emptyset$ since $\mathbf{a} \neq \mathbf{0}$. Let $\mathbf{e}_0 \in E$. Then $E = U + \mathbf{e}_0$. Hence, for any $\mathbf{w}_0 \in W$,

$$\begin{aligned}
\min_{\mathbf{e}\in E} \|\mathbf{w}_0 - \mathbf{e}\| &= \min_{\mathbf{e}\in E} \|(\mathbf{w}_0 - \mathbf{e}_0) - (\mathbf{e} - \mathbf{e}_0)\| \\
&= \min_{\mathbf{u}\in U} \|\mathbf{w}_0 - \mathbf{e}_0 - \mathbf{u}\| \\
&= \|(\mathbf{w}_0 - \mathbf{e}_0) - P_U(\mathbf{w}_0 - \mathbf{e}_0)\| \\
&= \frac{|\langle \mathbf{w}_0 - \mathbf{e}_0, \mathbf{a}\rangle|}{\|\mathbf{a}\|} \quad \text{(by (3.2) with } \mathbf{x} = \mathbf{w}_0 - \mathbf{e}_0\text{)} \\
&= \frac{|\langle \mathbf{w}_0, \mathbf{a}\rangle - c|}{\|\mathbf{a}\|}.
\end{aligned}$$

3.81 Let U and V be subspaces of a Euclidean space and assume that $U \cap V = \{\mathbf{0}\}$. Is it true that $U^\perp \cap V^\perp = \{\mathbf{0}\}$? Is it true that $(U+V)^\perp = U^\perp + V^\perp$?

Solution. No. Let $U = \text{Span}\,\{\mathbf{e}_1\}, V = \text{Span}\,\{\mathbf{e}_2\}$, where $\mathbf{e}_1 = (1,0,0), \mathbf{e}_2 = (0,1,0)$. Then $U \cap V = \{\mathbf{0}\}$ but $\mathbf{e}_3 \in U^\perp \cap V^\perp$. Also, $\dim(U+V) = 2$. Hence, $\dim(U+V)^\perp = 1$. But $\dim U^\perp = 2$, so $\dim(U^\perp + V^\perp)$ is at least 2. Hence, $(U+V)^\perp \neq U^\perp + V^\perp$.

3.83 Let $\mathbf{u}_1, \ldots, \mathbf{u}_m$ be distinct vectors in a Euclidean space X. Assume that $\mathbf{u}_i \perp \mathbf{u}_j$ for all $i \neq j$. Let $\mathbf{a} \in X$. Let S be the set of all numbers of the form

$$\left\| \mathbf{a} - \sum_{k=1}^{m} c_k \mathbf{u}_k \right\|,$$

where $c_1, \ldots, c_m$ range over all real numbers. Find $\inf S$.

Solution. Let U be the subspace spanned by $\mathbf{u}_i$s. We see that $\mathbf{u} \in U$ if and only if $\mathbf{u} = \sum_{k=1}^{m} c_k \mathbf{u}_k$. Hence $\inf S$ is also the $\inf_{\mathbf{u}\in U} \|\mathbf{a} - \mathbf{u}\|$. Theorem 3.5.12 shows that this infimum is $\|\mathbf{a} - P_U\mathbf{a}\| = \|P_{U^\perp}\mathbf{a}\|$. Here, if V is a subspace of X, then P_V is the orthogonal projection on V.

3.85 Let $E = \{\mathbf{e}_1, \ldots, \mathbf{e}_k\}$ and $U = \{\mathbf{u}_1, \ldots, \mathbf{u}_k\}$ be orthonormal subsets of a Euclidean space X such that Span E = Span U. Is it true that

$$\max_{\mathbf{x}\in X, \|\mathbf{x}\|=1} |\langle \mathbf{x}, \mathbf{e}_1\rangle|^2 + \cdots + |\langle \mathbf{x}, \mathbf{e}_k\rangle|^2 = \\ \max_{\mathbf{x}\in X, \|\mathbf{x}\|=1} |\langle \mathbf{x}, \mathbf{u}_1\rangle|^2 + \cdots + |\langle \mathbf{x}, \mathbf{u}_k\rangle|^2?$$

Solution. Yes. In fact, let $V = \text{Span}\, E = \text{Span}\, U$. We see that these maximums are $\max_{\mathbf{x}\in X, \|\mathbf{x}\|=1} \|P_V\mathbf{x}\|^2$. If V is not the zero-space then this maximum is one and otherwise it is zero.

3.87 Let $T : X \to X$ be an orthogonal projection. Is it true that

$$\langle T\mathbf{x}, \mathbf{y}\rangle = \langle \mathbf{x}, T\mathbf{y}\rangle \quad \text{for all } \mathbf{x}, \mathbf{y} \text{ in } X?$$

Solution. Yes. Let $U = TX$ and $V = U^{\perp}$ and let S be the orthogonal projection on V. Then $\langle T\mathbf{x}, \mathbf{y}\rangle = \langle T\mathbf{x}, T\mathbf{y} + S\mathbf{y}\rangle = \langle T\mathbf{x}, T\mathbf{y}\rangle$. The result follows by symmetry.

3.6 SPECTRAL THEOREM

3.89 Show that any orthogonal projection $T : X \to X$ to a subspace of X is a self-adjoint transformation.

Solution. This solution is the same as the solution to Problem 3.87.

3.91 Give an example of a self-adjoint transformation $T : X \to X$ and an eigenbasis $\mathbb{E}$ for T such that no vector in $\mathbb{E}$ is an eigenvector of T.

Solution. Define $T : \mathbb{R}^2 \to \mathbb{R}^2$ as $T(x, y) = (x, -y)$. We see that

$$\{(1, 1)/\sqrt{2}, (1, -1)/\sqrt{2}\}$$

is an eigenbasis for T. No vector in this basis is an eigenvector for T. A little extra work shows that this is essentially the only counterexample for this situation.

PART II

DIFFERENTIATION

CHAPTER 4

NORMED VECTOR SPACES

4.1 PRELIMINARIES

4.1 Let X be a vector space. If $\mathbf{a}, \mathbf{b}$ are in X, the line segment joining $\mathbf{a}$ and $\mathbf{b}$ is the set

$$L[\mathbf{a}, \mathbf{b}] = \{\, \mathbf{x} \in X \mid \mathbf{x} = t\mathbf{a} + (1-t)\mathbf{b},\ 0 \le t \le 1 \,\}.$$

Show that $L[\mathbf{a}, \mathbf{b}]$ is convex: If $\mathbf{u}, \mathbf{v}$ are in $L[\mathbf{a}, \mathbf{b}]$, then $s\mathbf{u} + (1-s)\mathbf{v} \in L[\mathbf{a}, \mathbf{b}]$ for all $0 \le s \le 1$.

Solution. Let $\mathbf{u}, \mathbf{v}$ be in $L[\mathbf{a}, \mathbf{b}]$. Then $\mathbf{u} = t_1\mathbf{a} + (1-t_1)\mathbf{b}, \mathbf{v} = t_2\mathbf{a} + (1-t_2)\mathbf{b}$ for some $0 \le t_1, t_2 \le 1$. Let $0 \le s \le 1$. Then

$$\begin{aligned} s\mathbf{u} + (1-s)\mathbf{v} &= s(t_1\mathbf{a} + (1-t_1)\mathbf{b}) + (1-s)(t_2\mathbf{a} + (1-t_2)\mathbf{b}) \\ &= (st_1 + (1-s)t_2)\mathbf{a} + (s(1-t_1) + (1-s)(1-t_2))\mathbf{b} \end{aligned}$$

Analysis in Vector Spaces.
By M. A. Akcoglu, P. F. A. Bartha and D. M. Ha

Let $c = st_1 + (1-s)t_2$. Since $0 \leq s \leq 1$ and t_1, t_2 are nonnegative, we have $0 \leq c$. Also, since t_1, t_2 are no more than 1, we have $c \leq s + (1-s) = 1$. Finally, it is easy to verify that $1 - c = s(1-t_1) + (1-s)(1-t_2)$. Thus,

$$s\mathbf{u} + (1-s)\mathbf{v} = c\mathbf{a} + (1-c)\mathbf{b} \in L[\mathbf{a}, \mathbf{b}].$$

So, $L[\mathbf{a}, \mathbf{b}]$ is convex.

4.3 Let $\| \ \|$ be the standard Euclidean norm on $\mathbb{R}^n$, and let $\|\mathbf{x}\|' = |x_1| + \cdots + |x_n|$ for all $\mathbf{x} \in \mathbb{R}^n$. Show that there are constants A, B such that

$$A\|\mathbf{x}\|' \leq \|\mathbf{x}\| \leq B\|\mathbf{x}\|' \quad \text{for all } \mathbf{x} \in \mathbb{R}^n.$$

(So these two norms are equivalent.)

Solution. Let $\mathbf{x} \in \mathbb{R}^n$. By the Cauchy-Schwartz inequality, we have

$$\|\mathbf{x}\|' = |x_1| + \cdots + |x_n| \leq (1^2 + \cdots + 1^2)^{1/2}(|x_1|^2 + \cdots + |x_n|^2)^{1/2} = \sqrt{n}\,\|\mathbf{x}\|.$$

Of course, we always have

$$\|\mathbf{x}\|^2 = x_1^2 + \cdots + x_n^2 \leq (|x_1| + \cdots + |x_n|)^2 = \|\mathbf{x}\|'^2.$$

Hence,

$$(1/\sqrt{n})\,\|\mathbf{x}\|' \leq \|\mathbf{x}\| \leq \|\mathbf{x}\|'.$$

4.5 Let $\| \ \|$ be a norm on X and let $\mathbf{x}, \mathbf{y}$ be in X. Assume that $\|3\mathbf{x} - \mathbf{y}\| \leq 1$ and $\|5\mathbf{x} - 4\mathbf{y}\| \leq 3$. Show that $\|\mathbf{y}\| \leq 2$.

Solution. We have $7\mathbf{y} = 5(3\mathbf{x} - \mathbf{y}) + 3(4\mathbf{y} - 5\mathbf{x})$ so that

$$7\|\mathbf{y}\| = \|7\mathbf{y}\| \leq 5\|3\mathbf{x} - \mathbf{y}\| + 3\|4\mathbf{y} - 5\mathbf{x}\| \leq 5 + 9 = 14.$$

4.7 Let M be a subspace of a normed space X. If there is some $\mathbf{a} \in X$ and some $r > 0$ such that $B_r(\mathbf{a}) \subset M$, show that $M = X$.

Solution. Suppose that $B_r(\mathbf{a}) \subset M$. Then $\mathbf{a} \in M$ and since M is a subspace, we have $B_r(\mathbf{0}) = -\mathbf{a} + B_r(\mathbf{a}) \subset -a + M = M$. If $\mathbf{x} \in X$, then by choosing $0 < c$ small enough, we have $\|c\mathbf{x}\| < r$ so that $c\mathbf{x} \in B_r(\mathbf{a})$. Hence, $c\mathbf{x} \in M$, and because M is a subspace, it follows that $\mathbf{x} \in M$. So, $X \subset M$.

4.9 Let X be a normed space. Let $\mathbf{a} \in X$ and $r > 0$. Show that

$$\frac{1}{r}\,(-\mathbf{a} + B_r(\mathbf{a})) = B_1(\mathbf{0}).$$

Solution. Note that $-\mathbf{a} + B_r(\mathbf{a}) = B_r(\mathbf{0})$ because

$$\mathbf{x} + \mathbf{a} \in B_r(\mathbf{a}) \iff \|(\mathbf{x} + \mathbf{a}) - \mathbf{a}\| < r \iff \|\mathbf{x} - \mathbf{0}\| < r \iff \mathbf{x} \in B_r(\mathbf{0}).$$

Similarly,

$$\mathbf{x} \in B_r(\mathbf{0}) \iff \|\mathbf{x}\| < r \iff \frac{1}{r}\|\mathbf{x}\| < 1 \iff \frac{1}{r}\mathbf{x} \in B_1(\mathbf{0}).$$

4.11 Let X be a vector space with $\dim X = 1$. Given a nonzero $\mathbf{e} \in X$ and a norm $\| \ \|$ on X, there is some positive constant C such that $\|a\mathbf{e}\| = |a|C$ for all $a \in \mathbb{R}$. True or false?

Solution. True. Suppose that $\dim X = 1$. Let $\mathbf{e} \in X$ with $\mathbf{e} \neq \mathbf{0}$ and let $\| \ \|$ be any norm on X. Let $C = \|\mathbf{e}\|$. Then $C > 0$ and for any $a \in \mathbb{R}$, we have $\|a\mathbf{e}\| = |a|\|\mathbf{e}\| = |a|\, C$.

4.13 Show that, for any real x, y, and z,

$$\begin{aligned}\sqrt{(x+2y)^2 + (y+2z)^2 + (z+2x)^2} &\leq \sqrt{(x-y)^2 + (y-z)^2 + (z-x)^2} \\ &+ 3\sqrt{x^2 + y^2 + z^2}.\end{aligned}$$

Solution. Apply the triangle inequality, stated as the inequality (4.1) in Remarks 4.1.8, with $\| \ \|$ as the standard Euclidean norm on $\mathbb{R}^3$ to

$$\begin{aligned}\mathbf{a} &= (2x + y, 2y + z, 2z + x), \\ \mathbf{b} &= (x + 2y, y + 2z, z + 2x), \\ \mathbf{c} &= (x - y, y - z, z - x).\end{aligned}$$

4.2 CONVERGENCE IN NORMED SPACES

4.15 Let $\mathbf{x}_n$ be a sequence in a normed space. Let $f : \mathbb{N} \to \mathbb{N}$ be an increasing function. If $\mathbf{x}_n$ converges, show that the sequence $\mathbf{x}_{f(n)}$ also converges.

Solution. Suppose that $\mathbf{x}_n \to \mathbf{a}$ as $n \to \infty$. Let $\epsilon > 0$ be given. Then there is some $N \in \mathbb{N}$ such that

$$\|\mathbf{x}_n - \mathbf{a}\| < \epsilon \quad \text{for all } n \geq N.$$

Since $f : \mathbb{N} \to \mathbb{N}$ is increasing, there is some M such that $f(M) \geq N$. If $k \geq M$, then $f(k) > f(M) \geq N$ so that

$$\left\|\mathbf{x}_{f(k)} - \mathbf{a}\right\| < \epsilon.$$

Hence, $(\mathbf{x}_{f(k)})$ converges to $\mathbf{a}$.

4.17 Define norms $\| \|$ and $\| \|'$ on $\mathbb{R}^2$ by

$$\|\mathbf{x}\| = |x_1| + |x_2|, \quad \|\mathbf{x}\|' = \max\{|x_1|, |x_2|\} \quad \text{for all } \mathbf{x} = (x_1, x_2) \in \mathbb{R}^2.$$

Show directly that $\| \|$ and $\| \|'$ are equivalent norms on $\mathbb{R}^2$.

Solution. Simply note that for all $(x_1, x_2) \in \mathbb{R}^2$, we have

$$\max(|x_1|, |x_2|) \leq |x_1| + |x_2| \leq 2\max(|x_1|, |x_2|).$$

4.19 Let X be an inner product space, and let $\| \|$ be the norm induced by the inner product. Suppose that $\mathbf{x}_n$ is an orthogonal sequence in X that converges. Find $\lim_{n\to\infty} \mathbf{x}_n$.

Solution. Suppose that $\mathbf{a} = \lim_{n\to\infty} \mathbf{x}_n$. Let $\epsilon > 0$ be arbitrary. Then there is some N such that $\|\mathbf{x}_n - \mathbf{a}\| < \epsilon$ for all $n \geq N$. Hence, if $n, m \geq N$, then $\|\mathbf{x}_n - \mathbf{x}_m\| = \|(\mathbf{x}_n - \mathbf{a}) + (\mathbf{a} - \mathbf{x}_m)\| \leq 2\epsilon$ for all $n, m \geq N$. But since $(\mathbf{x}_n)$ is an orthogonal sequence,

$$\|\mathbf{x}_n\|^2 + \|\mathbf{x}_m\|^2 = \|\mathbf{x}_n - \mathbf{x}_m\|^2 < 4\epsilon^2 \quad \text{for all } n, m \geq N.$$

In particular, this shows that $\|\mathbf{x}_n\| \to 0$ as $n \to \infty$. But $\|\mathbf{a}\| = \lim_{n\to\infty}$ so that $\|\mathbf{a}\| = 0$ and thus, $\mathbf{a} = \mathbf{0}$.

4.21 Suppose that $\mathbf{a}_n$ is a sequence in a normed space that converges to some $\mathbf{a}$. Show that

$$\lim_{n\to\infty} \frac{1}{n}(\mathbf{a}_1 + \cdots + \mathbf{a}_n) = \mathbf{a}.$$

Solution. Set $\mathbf{u}_n = \mathbf{a}_n - \mathbf{a}$ for all $n \in \mathbb{N}$. Then $\mathbf{u}_n \to \mathbf{0}$ and

$$\begin{aligned} \frac{1}{n}(\mathbf{a}_1 + \cdots + \mathbf{a}_n) - \mathbf{a} &= \frac{1}{n}(\mathbf{a}_1 + \cdots + \mathbf{a}_n - n\mathbf{a}) \\ &= \frac{1}{n}(\mathbf{u}_1 + \cdots + \mathbf{u}_n). \end{aligned}$$

Set

$$A_n = \frac{1}{n}(\mathbf{u}_1 + \cdots + \mathbf{u}_n), \quad n \in \mathbb{N}.$$

Let $\epsilon > 0$. Since $\mathbf{u}_n \to \mathbf{0}$, there is some $N \in \mathbb{N}$ such that $\|\mathbf{u}_k\| < \epsilon$ for all $k \geq N$. Let $C = \|\mathbf{u}_1 + \cdots + \mathbf{u}_N\|$. Find $M \in \mathbb{N}$ such that $C/M < \epsilon$. Then for all $n \geq M$,

$$\begin{aligned} \|A_n\| &\leq \frac{1}{n}\left(\|\mathbf{u}_1 + \cdots + \mathbf{u}_N\| + \sum\nolimits_{k=N+1}^{n} \|\mathbf{u}_k\|\right) \\ &\leq \frac{C}{n} + \frac{n-N}{n}\epsilon \\ &< \frac{C}{M} + \left(1 - \frac{N}{n}\right)\epsilon \\ &< 2\epsilon. \end{aligned}$$

Thus, $A_n \to \mathbf{0}$ as $n \to \infty$.

4.23 Let X be a vector space with bases $\{\mathbf{u}_1, \ldots, \mathbf{u}_m\}$ and $\{\mathbf{v}_1, \ldots, \mathbf{v}_m\}$. Suppose $S \subset X$ and assume that there is a constant M such that whenever $\mathbf{x} \in S$ with $\mathbf{x} = c_1\mathbf{u}_1 + \cdots + c_m\mathbf{u}_m$, then $\max_{1 \leq k \leq m} |c_k| \leq M$. Show that there is a constant C such that whenever $\mathbf{x} \in S$ and $\mathbf{x} = d_1\mathbf{v}_1 + \cdots + d_m\mathbf{v}_m$, then

$$|d_1| + \cdots + |d_m| \leq C.$$

Solution. Define norms $\|\ \|_1$ and $\|\ \|_2$ on X by

$$\begin{aligned} \|\mathbf{x}\|_1 &= \max_{1 \leq k \leq m} |c_k| \quad \text{if } \mathbf{x} = c_1\mathbf{u}_1 + \cdots + c_m\mathbf{u}_m \\ \|\mathbf{x}\|_2 &= |d_1| + \cdots + |d_m| \quad \text{if } \mathbf{x} = d_1\mathbf{v}_1 + \cdots + d_m\mathbf{v}_m. \end{aligned}$$

Since any two norms on X are equivalent, there is some constant A such that

$$\|\mathbf{x}\|_2 \leq A\|\mathbf{x}\|_1 \quad \text{for all } \mathbf{x} \in X.$$

Let $\mathbf{x} \in S$ with $\mathbf{x} = d_1\mathbf{v}_1 + \cdots + d_m\mathbf{v}_m$. By assumption, there is some M such that $\|\mathbf{x}\|_1 \leq M$. Hence, we also have $\|\mathbf{x}\|_2 \leq A\|\mathbf{x}\|_1 \leq AM$. Set$C = AM$. Then

$$|d_1| + \cdots + |d_m| = \|\mathbf{x}\|_2 \leq C.$$

4.25 Let $\mathbf{a}_k = (a_{k,0}, a_{k,1}, \ldots, a_{k,n})$ be a bounded sequence in $\mathbb{R}^{n+1}$. Show that there is an increasing sequence m_k in $\mathbb{N}$, and a polynomial p of degree no more than n such that

$$\lim_{k \to \infty} \max_{t \in [0,1]} |a_{m_k,0} + a_{m_k,1}t + \cdots + a_{m_k,n}t^n - p(t)| = 0.$$

Solution. Indeed, let X be the subspace of $C[0,1]$ consisting of all polynomials of degree no more than n. Then X is finite-dimensional, and each function

$$p_k(t) = a_{k,0} + a_{k,1}t + \cdots + a_{k,n}t^n, t \in [0,1]$$

is in X. Put the sup-norm $\| \|$ on X. Then since $\mathbf{a}_k = (a_{k,0}, a_{k,1}, \ldots, a_{k,n})$ is a bounded sequence in $\mathbb{R}^{n+1}$, we see that p_k is a bounded sequence in $(X, \| \|)$. Thus, by the Bolzano Weierstrass theorem, Theorem 4.2.16, a subsequence p_{m_k} of p_k converges to some $p \in X$. Thus, the desired conclusion follows.

4.3 NORMS OF LINEAR AND MULTILINEAR TRANSFORMATIONS

4.27 Let X be a nonzero normed space and let $T \in L(X, Y)$ with $T \neq \mathbf{0}$. What is $\{ \|T\mathbf{x}\| \mid \mathbf{x} \in X \}$?

Solution. Find some $\mathbf{e} \in X$ with $T\mathbf{e} \neq \mathbf{0}$. Let $a = \|T\mathbf{e}\|$. Then $a > 0$. If $r > 0$, then

$$\left\| T\left(\frac{r}{a}\mathbf{e}\right) \right\| = \frac{r}{a}\|T\mathbf{e}\| = r.$$

Hence, $\{ \|T\mathbf{x}\| \mid \mathbf{x} \in X \} = [0, \infty)$.

4.29 Let $T : (X, \| \|) \to Y$ be a linear map between normed spaces. Let $L : X \to X$ be an isomorphism such that $\|L\mathbf{x}\| = 1$ for all $\mathbf{x} \in X$ with $\|\mathbf{x}\| = 1$. Define a norm $\| \|'$ on X by $\|\mathbf{x}\|' = \|L\mathbf{x}\|$ for all $\mathbf{x} \in X$. If A is the norm of $T : (X, \| \|) \to Y$, show that the norm B of $T : (X, \| \|') \to Y$ satisfies $B \geq A$.

Solution. We are given that

$$A = \sup \{ \|T\mathbf{x}\| \mid \mathbf{x} \in X, \|\mathbf{x}\| = 1 \}.$$

Let B be the norm of $T : (X, \| \|') \to Y$. Then

$$\begin{aligned} B &= \sup \{ \|T\mathbf{x}\| \mid \mathbf{x} \in X, \|\mathbf{x}\|' = 1 \} \\ &= \sup \{ \|T\mathbf{x}\| \mid \mathbf{x} \in X, \|L\mathbf{x}\| = 1 \}. \end{aligned}$$

Thus, since $\|L\mathbf{x}\| = 1$ for all $\mathbf{x} \in X$ with $\|\mathbf{x}\| = 1$, we see that $B \geq A$.

4.31 For $k = 1, \ldots, n$, let $T_k : Y \to X_k$ be linear maps between normed spaces. Let $\| \|_k$ be the norm on X_k. Consider the norm $\| \|'$ on $\mathbf{X} = X_1 \times \cdots \times X_k$ given by

$$\|(\mathbf{x}_1, \ldots, \mathbf{x}_n)\|' = (\|\mathbf{x}_1\|_1^2 + \cdots + \|\mathbf{x}_n\|_n^2)^{\frac{1}{2}}.$$

Let $T : Y \to \mathbf{X}$ by $T\mathbf{y} = (T_1\mathbf{y}, \ldots, T_n\mathbf{y})$ for all $\mathbf{y} \in Y$. Show that T is linear. Is it true that

$$\|T\|^2 \leq \|T_1\|^2 + \cdots + \|T_n\|^2?$$

Solution. Suppose that $\mathbf{u}, \mathbf{v}$ are in Y. Then

$$\begin{aligned} T(\mathbf{u}+\mathbf{v}) &= (T_1(\mathbf{u}+\mathbf{v}), \ldots, T_n(\mathbf{u}+\mathbf{v})) \\ &= (T_1\mathbf{u}+T_1\mathbf{v}, \ldots, T_n\mathbf{u}+T_n\mathbf{v}) \\ &= (T_1\mathbf{u}, \ldots, T_n\mathbf{u}) + (T_1\mathbf{v}, \ldots, T_n\mathbf{v}) \\ &= T\mathbf{u}+T\mathbf{v}. \end{aligned}$$

Similarly, $T(c\mathbf{u}) = cT\mathbf{u}$ for all scalars c.

Now, let $\mathbf{y} \in Y$ with $\|\mathbf{y}\| = 1$. Then $\|T_k\mathbf{y}\|^2 = (\|T_k\mathbf{y}\|)^2 \leq (\|T_k\|\|\mathbf{y}\|)^2 = \|T_k\|^2$ for all $k = 1, \ldots, n$. Thus,

$$\begin{aligned} \|T\mathbf{y}\|'^2 &= \|T_1\mathbf{y}\|_1^2 + \cdots + \|T_n\mathbf{y}\|_n^2 \\ &\leq \|T_1\|^2 + \cdots + \|T_n\|^2. \end{aligned}$$

Hence, $\|T\|^2 \leq \|T_1\|^2 + \cdots + \|T_n\|^2$.

4.33 Let X, Y be normed spaces. For $\mathbf{x} \in X$, define $\widehat{\mathbf{x}} : L(X,Y) \to Y$ by $\widehat{\mathbf{x}}(T) = T(\mathbf{x})$ for all $T \in L(X,Y)$. Show that $\widehat{\mathbf{x}}$ is a bounded linear map with $\|\widehat{\mathbf{x}}\| \leq \|\mathbf{x}\|$.

Solution. Let $\mathbf{x} \in X$. Then it is easy to verify that $\widehat{\mathbf{x}}$ is linear. Let $T \in L(X,Y)$ with $\|T\| = 1$. Then

$$\|\widehat{\mathbf{x}}(T)\| = \|T\mathbf{x}\| \leq \|T\| \, \|\mathbf{x}\|.$$

Hence, $\widehat{\mathbf{x}}$ s bounded and $\|\widehat{\mathbf{x}}\| \leq \|\mathbf{x}\|$.

4.35 Let $T \in L(X,X)$ and suppose that there is some $S \in L(X,X)$ such that $S^2 = T$. Then $\sqrt{\|T\|} \leq \|S\|$. True or false?

Solution. True. If $S^2 = T$, then $\|T\| = \|S^2\| \leq \|S\|^2$ so that $\sqrt{\|T\|} \leq \|S\|$.

4.4 CONTINUITY IN NORMED SPACES

4.37 Let $f : X \to X$ be defined by $f(\mathbf{x}) = a\mathbf{x} + \mathbf{b}$, where X is a normed space, a is a constant, and $\mathbf{b}$ is a fixed vector in X. Show that f is uniformly continuous on X.

Solution. Let $\mathbf{u}, \mathbf{v}$ be in X. Then

$$\|f(\mathbf{u}) - f(\mathbf{v})\| = \|(a\mathbf{u}+\mathbf{b}) - (a\mathbf{v}+\mathbf{b})\| = |a| \, \|\mathbf{u}-\mathbf{v}\|.$$

So given any $\epsilon > 0$, if $\|\mathbf{u}-\mathbf{v}\| < \epsilon/(|a|+1)$, then $\|f(\mathbf{u}) - f(\mathbf{v})\| < \epsilon$. Hence, f is uniformly continuous.

4.39 Is there a continuous function $f : \mathbb{R}^3 \to \mathbb{R}$ with the following properties:

For each $k \in \mathbb{N}$, there are $\mathbf{x}_k, \mathbf{y}_k$ with $\|\mathbf{x}_k\| = \|\mathbf{y}_k\| = 1$ such that

$$\|\mathbf{x}_k - \mathbf{y}_k\| < (1/k) \quad \text{and} \quad f(\mathbf{x}_k) - f(\mathbf{y}_k) > 1?$$

Solution. No. Suppose that $f : \mathbb{R}^3 \to \mathbb{R}$ is continuous. Let there be sequences $(\mathbf{x}_k), (\mathbf{y}_k)$ with $\|\mathbf{x}_k\| = \|\mathbf{y}_k\| = 1$ and

$$\|\mathbf{x}_k - \mathbf{y}_k\| < \frac{1}{k}.$$

By the Bolzano Weierstrass theorem, Theorem 4.2.16, there is a subsequence $(\mathbf{x}_{k_l})$ of $(\mathbf{x}_k)$ that converges to some $\mathbf{z} \in \mathbb{R}^3$. Since $\|\mathbf{x}_k - \mathbf{y}_k\| \leq 1/k$ for all k, it follows that $(\mathbf{y}_{k_l})$ also converges to $\mathbf{z}$. Since f is continuous, it is continuous at $\mathbf{z}$. Hence, there is some $\delta > 0$ such that $|f(\mathbf{z}) - f(\mathbf{x})| < 1/4$ whenever $\mathbf{x} \in \mathbb{R}^3$ and $\|\mathbf{z} - \mathbf{x}\| < \delta$. Since $\mathbf{x}_{k_l} \to \mathbf{z}$ and $\mathbf{y}_{k_l} \to \mathbf{z}$, it follows that there is some N such that

$$|f(\mathbf{z}) - f(\mathbf{x}_{k_l}| < 1/4 \quad \text{and} \quad |f(\mathbf{z}) - f(\mathbf{y}_{k_l})| < 1/4 \quad \text{for all } l \geq N.$$

Hence, by the triangle inequality

$$|f(\mathbf{x}_{k_l}) - f(\mathbf{y}_{k_l})| \leq 1/2 \quad \text{for all } l \geq N.$$

4.41 Let $\| \|_1, \| \|_2$ be norms on a vector space X. Let $f : (X, \| \|_1) \to \mathbb{R}$ be defined by

$$f(\mathbf{x}) = \|\mathbf{x}\|_2 \quad \text{for all } \mathbf{x} \in X.$$

Show that f is uniformly continuous.

Solution. Since any two norms on X are equivalent, there are positive constants A and B such that

$$A\|\mathbf{x}\|_1 \leq \|\mathbf{x}\|_2 \leq B\|\mathbf{x}\|_1 \quad \text{for all } \mathbf{x} \in X.$$

Let $\epsilon > 0$ be given. Then whenever $\mathbf{u}, \mathbf{v}$ are in X and $\|\mathbf{u} - \mathbf{v}\|_1 < \delta$, we have

$$\begin{aligned}
|f(\mathbf{u}) - f(\mathbf{v})| &= | \|\mathbf{u}\|_2 - \|\mathbf{v}\|_2| \\
&\leq \|\mathbf{u} - \mathbf{v}\|_2 \\
&\leq B\|\mathbf{u} - \mathbf{v}\|_1 \\
&< B\epsilon.
\end{aligned}$$

Since $\epsilon > 0$ was arbitrary, the desired conclusion follows.

4.43 If $I - S \in L_0(X, X)$, must it follow that $\|S\| < 1$?

Solution. No. Define $S : \mathbb{R} \to \mathbb{R}$ as $Sx = 3x/2$, $x \in \mathbb{R}$. Then $(I - S)x = -x/2$. Hence $(I - S) : \mathbb{R} \to \mathbb{R}$ is invertible. But $\|S\| = 3/2 > 1$.

4.45 Suppose that $T_n \in L_0(X, X)$ for all $n \in N$ and $T_n \to T$. Must it be true that $T \in L_0(X, X)$?

Solution. No. Define $T_n \in L_0(\mathbb{R}, \mathbb{R})$ as $T_n x = x/n$, $x \in \mathbb{R}$. Then $T_n \to T$, where $Tx = 0$, $x \in \mathbb{R}$. Hence $T \notin L_0(\mathbb{R}, \mathbb{R})$.

4.47 Let X and Y be normed spaces. Let M be a nonempty compact subset of X. Let $\mathcal{C}(M, Y)$ be the collection of all continuous functions from M into $(Y, \|\ \|_Y)$. For f, g in $\mathcal{C}(M, Y)$ and scalar t, define $f + g$ and tf in the usual way: for all $\mathbf{m} \in M$,

$$\begin{aligned}(f + g)(\mathbf{m}) &= f(\mathbf{m}) + g(\mathbf{m}) \\ (tf)(\mathbf{m}) &= t(f(\mathbf{m})).\end{aligned}$$

Show that $\mathcal{C}(M, Y)$ is a vector space, which may or may not be finite-dimensional. Furthermore, show that the function

$$\|f\| = \max_{\mathbf{m} \in M} \|f(\mathbf{m})\|_Y \quad \text{for all } f \in \mathcal{C}(M, Y)$$

defines a norm on $\mathcal{C}(M, Y)$.

Solution. That $\mathcal{C}(M, Y)$ is a vector space follows directly from Example 3.1.1, as a special case of this example. To show that $\|f\| = \max_{\mathbf{m} \in M} \|f(\mathbf{m})\|_Y$ is a norm on this space first we must show the existence of this maximum. This would follow easily from the future Theorem 4.5.43, but let us prove this with the material we have so far as an exercise.

First we show that $A = \{ \|f(\mathbf{m})\|_Y \mid \mathbf{m} \in M \}$ is a bounded set in $\mathbb{R}$. If not there is a sequence $\mathbf{m}_n \in M$ such that $n < \|f(\mathbf{m}_n)\|$ for all $n \in \mathbb{N}$. Then, by Definition 4.4.19 of compactness, there is a subspace of $\mathbf{m}_n$ converging to a point in M. Without loss of generality assume that $\mathbf{m}_n \to \mathbf{a} \in M$. Then $\lim_n f(\mathbf{m}_n) = f(\mathbf{a})$ by the continuity of $f : M \to Y$. The reverse triangle inequality in Remarks 4.1.8 shows that $\lim_n \|f(\mathbf{m}_n)\| = \|f(\mathbf{a})\|$. This is a contradiction. Hence A is a bounded set in $\mathbb{R}$. In this case there is a sequence $\mathbf{x}_n \in M$ such that $\|f(\mathbf{x}_n)\|$ converges to $\sup_{r \in A} |r|$. By the compactness of M we assume, without loss of generality, that $\mathbf{x}_n$ converges to $\mathbf{c} \in M$. Then $\|f(\mathbf{x}_n)\|$ converges to $\|f(\mathbf{c})\| = \max_{\mathbf{m} \in M} \|f(\mathbf{m})\|$. This shows the existence of $\|\cdot\| : \mathcal{C}(M, Y) \to \mathbb{R}$. It is clear that this function satisfies the first two properties in the Definition 1.1.1 of norms. To obtain the the triangle inequality let f, $g \in \mathcal{C}(M, Y)$. Then

$$\begin{aligned}\|(f + g)(\mathbf{m})\|_Y &= \|f(\mathbf{m}) + g(\mathbf{m})\|_Y \\ &\leq \|f(\mathbf{m})\|_Y + \|g(\mathbf{m})\|_Y \leq \|f\|_{\mathcal{C}} + \|g\|_{\mathcal{C}}\end{aligned}$$

shows that $\|f + g\|_{\mathcal{C}} \leq \|f\|_{\mathcal{C}} + \|g\|_{\mathcal{C}}$.

4.5 TOPOLOGY OF NORMED SPACES

4.49 Show that if a closed set contains E then it also contains $\overline{E}$.

Solution. Let F be a closed set containing E. Let $\mathbf{a} \in \overline{E}$. Theorem 4.5.27 shows that there is a sequence $\mathbf{a}_n$ in E that converges to $\mathbf{a}$. Then $\mathbf{a}_n$ is also a convergent sequence in F. Then Theorem 4.5.25 shows that F contains the limit point $\mathbf{a}$ since F is a closed set. Hence $\mathbf{a} \in F$ and, therefore, $\overline{E} \subset F$.

4.51 Let A, B be subsets of a normed space. Show that

$$\partial(A \cup B) \subset \partial A \cup \partial B.$$

Give an example to show that it is possible to have $\partial(A \cup B) \neq \partial A \cup \partial B$. If $A \subset B$, must it follow that $\partial A \subset \partial B$?

Solution. Let $\mathbf{x} \in \partial(A \cup B)$ and let $r > 0$. Then $B_r(\mathbf{x}) \cap (A \cup B) \neq \emptyset$ and $B_r(\mathbf{x}) \cap (A \cup B)^c \neq \emptyset$. Since $(A \cup B)^c = A^c \cap B^c$, it follows that $B_r(\mathbf{x}) \cap A^c \neq \emptyset$ and $B_r(\mathbf{x}) \cap B^c \neq \emptyset$. Since $B_r(\mathbf{x}) \cap (A \cup B) \neq \emptyset$, we must have $B_r(\mathbf{x}) \cap A \neq \emptyset$ or $B_r(\mathbf{x}) \cap B \neq \emptyset$. Thus, $\mathbf{x} \in \partial A$ or $\mathbf{x} \in \partial B$. Hence, $\mathbf{x} \in \partial A \cup \partial B$.

Let $A = [0, 1], B = [1, 2]$. Then $\partial(A \cup B) = \partial([0\,2]) = \{0, 2\}$ but $\partial A \cup \partial B = \{0, 1\} \cup \{1, 2\} = \{0, 1, 2\}$.

If $A \subset B$, then it is not necessarily true that $\partial A \subset \partial B$. For example, take $A = \{1\}$ and $B = \mathbb{R}$. Then $\partial A = A$ but $\partial B = \emptyset$.

4.53 Given a normed space X, find an example of a set E such that $\partial E = X$.

Solution. Let $\{\, \mathbf{u}_1, \ldots, \mathbf{u}_k \}$ be a basis for X. Let E be the set of all points $\mathbf{x} = \sum_i x_i \mathbf{u}_i$ where all x_is are rational numbers. We claim that $\partial E = X$. In fact, given any real number x there is a sequence of rational numbers r_n and a sequence of irrational numbers s_n both converging to x. Then, given any $\mathbf{x} \in X$ there is a sequence $\mathbf{r}_n$ in E and a sequence $\mathbf{s}_n$ in E^c both converging to $\mathbf{x}$. Hence $\mathbf{x} \in \partial E$.

4.55 Let $\mathbf{x}_n$ be a sequence in a normed space X and assume that $\mathbf{x}_n \to \mathbf{a}$ for some $\mathbf{a} \in X$. Let $S = \{\, \mathbf{x}_n \mid n \in \mathbb{N} \,\}$. Is it true that $\partial S = \{\mathbf{a}\}$?

Solution. If $\mathbf{x}_n = \mathbf{a}$ for all $n \in \mathbb{N}$ then we see that $\partial S = \{\mathbf{a}\}$. We also see that each $\mathbf{x}_n$ is a boundary point of S. Hence it is not true that $\partial S = \{\mathbf{a}\}$, unless $\mathbf{x}_n = \mathbf{a}$ for all $n \in \mathbb{N}$.

4.57 Let X be a normed space. Verify that $\partial\emptyset = \emptyset = \partial X$. If $E \subset X$, must it be true that $\partial(\partial E) \subset \partial E$?

Solution. For any $\mathbf{x} \in X$, we have $B_r(\mathbf{x}) \cap \emptyset = \emptyset$. Hence, $\mathbf{x} \notin \partial\emptyset$ and $\mathbf{x} \notin \partial X$. Hence, no $\mathbf{x}$ in X can be a boundary point of $\emptyset$ or of X. Thus, $\partial\emptyset = \emptyset = \partial X$.

For the second part, let $\mathbf{a} \in \partial(\partial E)$. Then any neighborhood G of $\mathbf{a}$ intersects ∂E. Hence G is also a neighborhood of a point in ∂E. Therefore G intersects both E and E^c. Hence $\mathbf{a} \in \partial E$ and $\partial(\partial E) \subset \partial E$.

4.59 Let F be any finite subset of a nonzero normed space X. Show that $\partial F = F$. Hence, deduce that F is closed.

Solution. If $F = \emptyset$, then no point in X can be a boundary point of F. Thus, in this case, $\partial F = \emptyset = F$. Suppose that $F \neq \emptyset$, say, $F = \{\mathbf{x}_1, \ldots, \mathbf{x}_m\}$. If $r > 0$, then $B_r(\mathbf{x}_i)$ is an infinite subset of X, and thus, it must contain some point in F^c. Of course, $B_r(\mathbf{x}_i)$ contains $\mathbf{x}_i$. So, each $\mathbf{x}_i$ is a boundary point of F. Thus, $F \subset \partial F$.

On the other hand, let $\mathbf{y} \in F^c$. Since F is finite,

$$r = \min \{ \|\mathbf{y} - \mathbf{x}_k\| \mid 1 \leq k \leq m \} > 0.$$

Thus, $B_\eta(\mathbf{y})$ contains no points in F whenever $0 < \eta < r$. Hence, $\mathbf{y}$ cannot be a boundary point of F. Hence, $\partial F \subset F$. Thus, $F = \partial F$, as to be shown.

4.61 Let F be a finite subset of an open set O in a normed space. Show that $O \setminus F$ is also open.

Solution. Every finite subset of a normed space is closed. Hence, F^c is open. Thus, $O \setminus F = O \cap F^c$ is an intersection of open sets. Thus, it must be open.

4.63 Show that an arbitrary intersection of closed sets is closed and a union of finitely many closed sets is also closed.

Solution. Let F_α, $\alpha \in A$, be a family of closed sets and $F = \cap_\alpha F_\alpha$. Let $\mathbf{x}_n$ be a sequence in F that converges to a point $\mathbf{x}$. Then $\mathbf{x}_n$ is a convergent sequence in each F_α. Since F_α is a closed set Theorem 4.5.25 shows that $\mathbf{x} \in F_\alpha$ for all $\alpha \in A$. Therefore $\mathbf{x} \in F = \cap_\alpha F_\alpha$. Hence F contains the limit of every convergent sequence in F. Then Theorem 4.5.25 shows that F is closed.

Now let $F = \cup_i F_i$ be a finite union of closed sets F_i. Let x_n be a sequence in F that converges to a point $\mathbf{x}$. Now this sequence is contained in the union of sets F_i. Then we verify easily that at least one F_{i_0} of these sets must contained an infinite number of terms from this sequence. Hence F_{i_0} contains a subsequence of this convergent sequence, which also converges to $\mathbf{x}$. Since F_{i_0} is a closed set, Theorem 4.5.25 shows that $\mathbf{x} \in F_{\alpha_0}$. Therefore $\mathbf{x} \in F = \cup_i F_i$. Then Theorem 4.5.25 shows that F is closed.

4.65 Let $U = \{ (x, y, z) \in \mathbb{R}^3 \mid |2x + 3y + z| < 1, |x - y + 5z| < 3 \}$. Show that U is open in the Euclidean space $\mathbb{R}^3$.

Solution. Let $\varphi(x, y, z) = 2x + 3y + z$. Then $\varphi : \mathbb{R}^3 \to \mathbb{R}$ is a continuous function. This follows from the linearity of φ and from Theorem 4.4.8 that shows that linear functions are continuous. Then

$$U_1 = \{ (x, y, z) \in \mathbb{R}^3 \mid |x - y + 5z| < 3 \} = \varphi^{-1}(-1, 1)$$

is an open set in X, as the inverse image of the open set $(-1, 1) \subset \mathbb{R}$ under the continuous function $\varphi : \mathbb{R}^3 \to \mathbb{R}$. This follows from Corollary 4.5.30. Similarly

$$U_2 = \{ (x, y, z) \in \mathbb{R}^3 \mid |2x + 3y + z| < 1 \}$$

is also an open set. Then $U = U_1 \cap U_2$ is open as the intersection of two open sets. This follows from Theorem 4.5.5.

4.67 Suppose that X and Y are normed spaces and let $f : X \to Y$ be continuous. Let $D \subset X$. If f is one-to-one on X, show that $f(\partial D) \subset \partial(f(D))$. What happens if f is not one-to-one?

Solution. Let $\mathbf{b} \in f(\partial D)$. Since f is one-to-one there is a unique point $\mathbf{a} \in \partial D$ such that $f(\mathbf{a}) = \mathbf{b}$. Let $H \subset Y$ be a neighborhood of $\mathbf{b}$. Corollary 4.5.30 shows that $G = f^{-1}(H)$ is an open set in X. Also, $\mathbf{a} \in G \cap \partial D$. Hence G is a neighborhood of $\mathbf{a} \in \partial D$. Therefore G intersects both D and D^c. Since f is one-to-one $H = f(G)$ intersects both $f(D)$ and $f(D^c) = (f(D))^c$. Hence $\mathbf{b} \in \partial f(D)$ and $f(\partial D) \subset \partial f(D)$.

This may not be true if f is not one-to-one. Define $f : \mathbb{R}^2 \to \mathbb{R}$ by $f(x, y) = x$. Let

$$D = \{ (x, 0) \mid x \in \mathbb{R} \} \subset \mathbb{R}^2.$$

We see that $\partial D = D$ and $f(D) = f(\partial D) = \mathbb{R}$. But $\partial f(D) = \partial \mathbb{R} = \emptyset$ in $\mathbb{R}$.

4.69 Let E be a set in a normed space X. A point $\mathbf{a} \in X$ is an *accumulation point of* E if $E \cap B_r(\mathbf{a})$ is an infinite set (that is, contains infinitely many points) for all $r > 0$. Show that every bounded infinite set has an accumulation point. Is every point in ∂E an accumulation point of E?

Solution. Let $\mathbf{x}_n$ be a convergent sequence consisting of all different terms, that is, $\mathbf{x}_i \neq \mathbf{x}_j$ whenever $i \neq j$. Then $\mathbf{x}$ is an accumulation point of

$$E = \{ \mathbf{x}_n \mid n \in \mathbb{N} \},$$

as we see by an easy verification. Bolzano-Weierstrass theorem, Theorem 4.2.16, shows that any bounded infinite set contains a convergent sequence consisting of different terms. Hence any bounded infinite set has an accumulation point.

A point in ∂E need not be an accumulation point of E. In fact a single-point set $E = \{\mathbf{a}\}$ has a boundary point but no accumulation points.

Problems on Compact Sets

4.71 Let A, B be compact subsets of a normed space. Show that $A \cap B$ and $A \cup B$ are compact.

Solution. Both A and B are bounded and closed sets. This follows from Theorem 4.5.38. Then $A \cap B$ and $A \cup B$ are bounded sets. They are also closed by Theorem 4.5.5 and by the definition of closed sets, Definition 4.5.16. Hence $A \cap B$ and $A \cup B$ are compact, again by Theorem 4.5.38.

4.73 Let X be a normed space. Let A be a compact subset of X, and let B be a nonempty finite subset of X. Must $A + B = \{\, \mathbf{a} + \mathbf{b} \mid \mathbf{a} \in A,\ \mathbf{b} \in B \,\}$ be compact?

Solution. Define the translation by $\mathbf{b} \in X$ as the function

$$T_{\mathbf{a}}(\mathbf{x}) = \mathbf{b} + \mathbf{x}, \quad \mathbf{x} \in X.$$

. We see that $T_{\mathbf{b}} : X \to X$ is a continuous function. Hence

$$T_{\mathbf{b}}(A) = \{\, \mathbf{a} + \mathbf{b} \mid \mathbf{a} \in A \,\}$$

is a compact set. This follows from Theorem 4.5.43. Then, by Problem 4.71,

$$A + B = \{\, \mathbf{a} + \mathbf{b} \mid \mathbf{a} \in A,\ \mathbf{b} \in B \,\} = \bigcup\nolimits_{\mathbf{b} \in B} T_{\mathbf{b}}(A)$$

is a compact set as a finite union of compact sets.

4.75 Let E be a subset of a normed space X. If $\mathbf{a} \in X$ is an accumulation point (Problem 4.69) of E then $E \setminus \{\mathbf{a}\}$ is not compact. If $\mathbf{u} \in \partial E$, is it true that $E \setminus \{\mathbf{u}\}$ is not compact?

Solution. Let $\mathbf{a}$ be an accumulation point of E. Let $A = E \setminus \{\mathbf{a}\}$. Then we see that $A \cap B_r(\mathbf{a})$ is a nonempty set for all $r > 0$. Choose $\mathbf{x}_n \in A \cap B_{1/n}(\mathbf{a})$ for each $n \in \mathbb{N}$. Then $\mathbf{x}_n$ is a sequence in A converging to $\mathbf{a} \notin A$. Hence A is not closed, by Theorem 4.5.25, and therefore not compact, by Theorem 4.5.38. For a general set E, if $\mathbf{u} \in \partial E$ then $E \setminus \{\mathbf{u}\}$ may or may not be compact. For the first possibility let $E = \{\mathbf{u}, \mathbf{v}\}$ be a two-point set, with $\mathbf{u} \neq \mathbf{v}$. Then $E = \partial E$ and $E \setminus \{\mathbf{u}\} = \{\mathbf{v}\}$ is compact. For the second possibility let E be a set that has an accumulation point $\mathbf{u}$. Then $\mathbf{u} \in \partial E$ and $E \setminus \{\mathbf{a}\}$ is not compact, as in the first part of this solution.

4.77 Let E_n be a sequence of nonempty compact subsets of a normed space X. If $E_{k+1} \subset E_k$ for all $k \in \mathbb{N}$, show that $\cap_n E_n \neq \emptyset$. Give an example of a sequence A_n of nonempty subsets of $\mathbb{R}$ with $A_{k+1} \subset A_k$ for all $k \in \mathbb{N}$, but such that $\cap_n A_n = \emptyset$.

Solution. Choose point $\mathbf{x}_n$ in each E_n. Then $\mathbf{x}_n$ is a sequence in a compact set E_1. Therefore a subspace converges to a point $\mathbf{x}$. We see that each E_n contains all the terms of this subspace with indices greater than n. But each E_n is compact. Therefore each E_n contains the limit point $\mathbf{x}$. Hence $\mathbf{x} \in \cap_n E_n$ and therefore $\cap_n E_n \neq \emptyset$. As an example for the second part take $A_n = (0, 1/n)$.

4.79 Let X and Y be normed spaces. Let M be a compact set in X. Let $f : M \to Y$ be a continuous and one-to-one function. Show that a sequence $\mathbf{m}_n \in M$ converges if and only if $f(\mathbf{m}_n) \in Y$ converges.

Solution. Let $L = f(M)$. This is a compact set in Y, by Theorem 4.5.43. Also, $f : M \to L$ has an inverse function $g : L \to M$, since it is one-to-one on M. Theorem 4.4.21 shows that the inverse function $g : L \to M$ is also continuous. If $\mathbf{m}_n \in M$ converges in X then it converges to a point $\mathbf{m} \in M$, by the compactness of M. Then by the continuity of f at $\mathbf{m}$, Definition 4.4.2, $f(\mathbf{m}_n)$ converges in Y. Conversely, if $f(\mathbf{m}_n) = \mathbf{y}_n \in L$ converges in Y then it converges to a point $\mathbf{y} \in L$, by the compactness of L. Then by the continuity of g at $\mathbf{y}$, Definition 4.4.2, $g(\mathbf{y}_n) = \mathbf{m}_n$ converges in X.

Problems on Connected Sets

4.81 Let $C = A \cup B$, where A and B are two sets in the xy-plane defined by

$$A = \{\, (0, y) \mid -1 < y < 1 \,\} \text{ and } B = \{\, (x, \sin(1/x)) \mid 0 < x < 1 \,\}.$$

Show that C is connected but not arcwise connected.

Solution. Let P and Q be two nonempty sets such that $P \cup Q = C$. We will show that $C \cap \overline{P} \cap \overline{Q}$ is not empty. Then, by Definition 4.5.32, the connectedness of C will follow.

If A intersects both P and Q then $A \cap \overline{P} \cap \overline{Q}$ is not empty, since A is a connected set. This follows easily from Theorem 2.6.13 that shows that intervals are connected sets. If B intersects both P and Q then, again, $B \cap \overline{P} \cap \overline{Q}$ is not empty. To see this note that the sets

$$\begin{aligned} P' &= \{x \mid 0 < x < 1,\ (x, \sin(1/x)) \in P\} \text{ and} \\ Q' &= \{x \mid 0 < x < 1,\ (x, \sin(1/x)) \in Q\} \end{aligned}$$

are both nonempty and $P' \cup Q' = (0, 1)$. Hence there is an $s \in (0, 1) \cap \overline{P'} \cap \overline{Q'}$. In this case we see that $(s, \sin(1/s)) \in B \cap \overline{P} \cap \overline{Q}$.

Now assume that $P = A$ and $Q = B$. We see that $\overline{P} = \overline{A}$ is the vertical interval

$$I = \{\, (0, y) \mid -1 \leq y \leq 1 \,\}$$

on the y-axis. We claim that $I \subset \overline{Q}$. In fact, if $(0, a) \in I$ then B intersects the horizontal line $y = a$ at the points (x_n, a) with $\sin(1/x_n) = a$, or $x_n =$

$1/(\sin^{-1} a + 2\pi n)$. We see that $(x_n, a) \to (0, a)$. Hence $I \subset \overline{B}$. Therefore $C \cap \overline{P} \cap \overline{Q} = I$ is not empty. Hence C is connected.

To see that C is not arcwise connected, let $\mathbf{r} : [0, 1] \to C$ be a continuous function. Assume that $\mathbf{r}(0) \in A$ and $\mathbf{r}(1) \in B$. We show that this leads to a contradiction. Let

$$U = \{\, t \mid \mathbf{r}(t) \in A \,\} \quad \text{and} \quad V = \{\, t \mid \mathbf{r}(t) \in B \,\}.$$

Then U and V are both nonempty and $U \cup V = [\,0,\ 1\,]$. Since $[\,0,\ 1\,]$ is a connected set $[\,0,\ 1\,] \cap \overline{U} \cap \overline{V}$ is nonempty. Hence we see that there are convergent sequences u_n in U and v_n in V converging to the same limit. Hence $\mathbf{r}(u_n) \in A$ and $\mathbf{r}(v_n) \in B$ converge to the same limit. This is a contradiction. In fact $\mathbf{r}(u_n) = (0,\ y_n)$ and $\mathbf{r}(s_n) = (x_n,\ \sin(1/x_n))$. If they converge to the same limit then x_n must converge to zero. In this case $\sin(1/x_n)$ cannot converge. Hence C is not arcwise connected.

4.83 Show that if an open set is connected, then it is also arcwise connected. (*Hint:* Let G be an open set and $\mathbf{a} \in G$. Let A be the set of all points in G that can be joined to $\mathbf{a}$ by an arc in C. Let $B = G \setminus A$. Show that A and B are both open.)

Solution. We follow the hint. Let $\mathbf{b} \in A$. Since $A \subset G$ and since G is open there is an $r > 0$ such that $B_r(\mathbf{b}) \subset G$. We claim that $B_r(\mathbf{b}) \subset A$. Let $\mathbf{c} \in B_r(\mathbf{b})$. There is a continuous function $\mathbf{r}_1 : [\,0,\ 1\,] \to G$ such that $\mathbf{r}_1(0) = \mathbf{a}$ and $\mathbf{r}_1(1) = \mathbf{b}$. Define $\mathbf{r}_2 : [\,1,\ 2\,] \to G$ as $\mathbf{r}_2(t) = (2 - t)\mathbf{b} + (1 - t)\mathbf{c}$. Then obtain $\mathbf{r} : [\,0,\ 2\,] \to G$ as

$$\mathbf{r}(t) = \begin{cases} \mathbf{r}_1(t) & \text{if } 0 \le t < 1 \text{ and} \\ \mathbf{r}_2(t) & \text{if } 1 \le t \le 2. \end{cases}$$

We see that $\mathbf{r} : [\,0,\ 2\,]$ is continuous function joining $\mathbf{a}$ to $\mathbf{c}$. Hence $\mathbf{c} \in A$. Therefore $B_r(\mathbf{b}) \subset A$ and A is open.

Similarly, $B = G \setminus A$ is also open. Let $\mathbf{b} \in B$ and $B_r(\mathbf{b}) \subset G$. We claim that if $\mathbf{c} \in B_r(\mathbf{b})$ then $\mathbf{c} \in B$. Otherwise $\mathbf{c}$ could be joined to $\mathbf{a}$ by an arc in G. In this case we see, as before, $\mathbf{b}$ could be also joined to $\mathbf{a}$ by an arc in G. This contradicts the assumption that $\mathbf{b} \in B = G \setminus A$.

To complete the proof we will show that $B = \emptyset$. First, $A \neq \emptyset$, since $\mathbf{a} \in A$. If B is also nonempty then there is an $\mathbf{x} \in G \cap \overline{A} \cap \overline{B}$, by the connectedness of G, Definition 4.5.32. Then either $\mathbf{x} \in A$ or $\mathbf{x} \in B$. In the first case $\mathbf{x} \notin \overline{B}$, since $\mathbf{x}$ is contained in the open set A disjoint from B. In the second case $\mathbf{x} \notin \overline{A}$. Hence both cases are impossible. Hence B must be empty and $A = G$. Any two points in G can be connected by a continuous arc in G.

4.85 Show that any open set in $\mathbb{R}$ is the union of a sequence of pairwise disjoint open intervals. Give an example to show that this union need not be a finite union, even for bounded open sets.

Solution. Let r_n be a dense sequence in $\mathbb{R}$. An example of such a sequence is the set of binary numbers, $k\,2^{-n}$ with $k \in \mathbb{Z}$ and $n \in \mathbb{N}$. We see that this set can be arranged as a sequence. Also note that if a finite number of terms is removed from a dense sequence the remaining sequence is also dense.

Take the first term r_{n_1} in a fixed dense sequence that is contained in the given open set A. Such a term exists by the definition of dense sequences. Let A_1 be the set of points in A that can be connected to r_{n_1} by a continuous arc in A. Solution to Problem 4.83 shows that A_1 and $B_1 = A \setminus A_1$ are disjoint open sets. Assume that the points $r_{n_1}, \ldots, r_{n_k}$ and the sets $A_1, \ldots, A_k$ are already defined as above so that A_is are pairwise disjoint open subsets of A such that $B_k = A \setminus \cup_{i=1}^{k} A_i$ is also open. If $B_k = \emptyset$ then we are done. Otherwise let $r_{n_{k+1}}$ be the first term of r_i that comes after r_{n_k} and is contained in B_k. Again, such a term exists by the definition of dense sequences. Let A_{k+1} be the set of all points in B_k that can be connected to $r_{n_{k+1}}$ by a continuous arc in B_k. This defines the induction step. We see that A_ks are pairwise disjoint connected open sets. Also, $A = \cup_k A_k$. In fact, if $x \in A$, then $B_s(x) \subset A$ for some $s > 0$. Hence $B_s(x)$ contains a term of the sequence r_n with the smallest index n_0. In this case we see that $x \in A_n$ with an $n \leq n_0$. Since the only open and connected sets of $\mathbb{R}$ are the open intervals each A_k is an open interval and A is the union of a sequence of pairwise disjoint open intervals.

As a simple example to show that $A = \cup A_k$ can be an infinite union let

$$A = \bigcup_{n \in \mathbb{N}} (1/(2n+1),\, 1/(2n)).$$

Even a simpler example is $A = (0,\, 1) \setminus \{1,\, 1/2,\, 1/3,\, 1/4,\, \ldots\}$.

Problems on Distances Between Sets

4.87 For any nonempty set E and for any $r > 0$ let $E_r = \cup_{\mathbf{x} \in E} B_r(\mathbf{x})$ be the *enlargement of E* by $r > 0$. Show that $\mathbf{x} \in E_r$ if and only if $\rho(\mathbf{x},\, E) < r$. Here

$$\rho(\mathbf{x},\, E) = \inf_{\mathbf{e} \in E} \|\mathbf{x} - \mathbf{e}\|$$

is the distance of the point $\mathbf{x}$ to the set E, as defined in Problem 4.86.

Solution. Let $\mathbf{x} \in E_r = \cup_{\mathbf{e} \in E} B_r(\mathbf{e})$. Then there is an $\mathbf{e} \in E$ such that $\mathbf{x} \in B_r(\mathbf{e})$. Hence $\|\mathbf{x} - \mathbf{e}\| < r$ and $\rho(\mathbf{x},\, E) < r$. Conversely, assume that $\rho(\mathbf{x},\, E) < r$. In this case there is an $\mathbf{e} \in E$ such that $\|\mathbf{x} - \mathbf{e}\| < r$. Hence $\mathbf{x} \in B_r(\mathbf{e})$ and $\mathbf{x} \in E_r$.

4.89 Let A and B be two nonempty sets in a normed space X. Let $\rho(\mathbf{x},\, B)$, $\mathbf{x} \in X$, be as defined in Problem 4.87. Let $\rho(A,\, B)$ be the distance between A and B, as defined in Definition 4.5.47. Show that $\rho(A,\, B) = \inf_{\mathbf{x} \in A} \rho(\mathbf{x},\, B)$.

Solution. We have $\rho(A,\, B) = \inf_{\mathbf{x} \in A,\, \mathbf{y} \in B} \|\mathbf{x} - \mathbf{y}\|$. Let $\mathbf{x} \in A$. Then

$$\rho(A,\, B) \leq \|\mathbf{x} - \mathbf{y}\| \quad \text{for all } \mathbf{y} \in B.$$

Hence $\rho(A, B) \leq \inf_{\mathbf{y}\in B} \|\mathbf{x}-\mathbf{y}\| = \rho(\mathbf{x}, B)$ for all $\mathbf{x} \in A$. Therefore

$$\rho(A, B) \leq \inf_{\mathbf{x}\in A}\rho(\mathbf{x}, B).$$

For the other direction, given $\varepsilon > 0$ find $\mathbf{a} \in A$, $\mathbf{b} \in B$ such that

$$\|\mathbf{a}-\mathbf{b}\| \leq \rho(A, B) + \varepsilon.$$

Hence $\rho(\mathbf{a}, B) = \inf_{\mathbf{y}\in B} \|\mathbf{a}-\mathbf{y}\| \leq \rho(A, B) + \varepsilon$. Therefore

$$\inf_{\mathbf{a}\in A}\rho(\mathbf{a}, B) \leq \rho(A, B) + \varepsilon.$$

This means that $\inf_{\mathbf{a}\in A}\rho(\mathbf{a}, B) \leq \rho(A, B)$. Hence $\inf_{\mathbf{a}\in A}\rho(\mathbf{a}, B) = \rho(A, B)$.

4.91 Let A and B be two nonempty sets in a normed space X. Give an example to show that there may not be any points $\mathbf{a} \in \overline{A}$ and $\mathbf{b} \in \overline{B}$ such that $\rho(A, B) = \|\mathbf{a}-\mathbf{b}\|$.

Solution. Let $A = \{\, n \mid n \in \mathbb{N} \,\}$ and $B = \{\, n + 1/n \mid n \in \mathbb{N} \,\}$, as subsets of $\mathbb{R}$. We see that $\rho(A, B) = 0$, since $\inf_n |n - (n + 1/n)| = \inf_n |1/n| = 0$. But both A and B are closed sets and they have no common points.

Problems on Convex Sets

4.93 Show that if a convex set in a normed space X contains points from a set A and from its complement $A^c = X \setminus A$, then it also contains points from ∂A. Give an example to show that this is not necessarily true for non-convex sets.

Solution. Let C be a convex set. Let $\mathbf{a} \in C \cap A$ and $\mathbf{b} \in C \cap A^c$. Then the segment

$$S = \{\, \mathbf{c}_t = (1-t)\mathbf{a} + t\mathbf{b} \mid 0 \leq t \leq 1 \,\}$$

joining these two points is contained in C, since C is convex. Let

$$U = \{\, t \mid t \in [\,0,\, 1\,],\ \mathbf{c}_t \in A \,\} \quad \text{and} \quad V = \{\, t \mid t \in [\,0,\, 1\,],\ \mathbf{c}_t \in A^c \,\}.$$

Both U and V are nonempty and $[\,0,\, 1\,] = U \cup V$. Therefore, since $[\,0,\, 1\,]$ is convex, there is an $\mathbf{a} \in [\,0,\, 1\,] \cap \overline{U} \cap \overline{V}$. Then we see that $\mathbf{c}_\alpha \in C \cap \partial A$.

This is not necessarily true for non-convex sets. Let, for example,

$$X = \mathbb{R}, \quad A = (0,\, 3), \quad \text{and} \quad C = (1,\, 2) \cup (4,\, 5).$$

Then $C \cap A$ and $C \cap A^c$ are nonempty but $C \cap \partial A = \emptyset$.

Problems on Oscillations

4.95 Let X and Y be normed spaces. Let $E \subset X$ and let $f : E \to Y$ be a bounded function. If $G \subset E$, then show that

$$\Omega(f,\, G) = \sup\{\, \|f(\mathbf{u}) - f(\mathbf{v})\| \mid \mathbf{u},\, \mathbf{v} \in G \,\}$$

exists. It is called the *oscillation of f over the set G*. Also show that if $A \subset B \subset E$ then $\Omega(f,\, A) \leq \Omega(f,\, B)$.

Solution. Let $\|f(\mathbf{u})\| \leq M$ for all $\mathbf{u} \in E$. Then $\|f(\mathbf{u}) - f(\mathbf{v})\| \leq 2M$ for all $\mathbf{u},\, \mathbf{v} \in M$. Hence $\{\, \|f(\mathbf{u}) - f(\mathbf{v})\| \mid \mathbf{u},\, \mathbf{v} \in G \,\}$ is a bounded set in $\mathbb{R}$. Therefore its supremum $\Omega(f,\, G)$ exists. Also, if $A \subset B \subset E$ then

$$\{\, \|f(\mathbf{u}) - f(\mathbf{v})\| \mid \mathbf{u},\, \mathbf{v} \in A \,\} \subset \{\, \|f(\mathbf{u}) - f(\mathbf{v})\| \mid \mathbf{u},\, \mathbf{v} \in B \,\} \subset \mathbb{R}.$$

Hence $\Omega(f,\, A) \leq \Omega(f,\, B)$.

4.97 Show that f is continuous at $\mathbf{a}$ if and only if $\omega(f,\, \mathbf{a}) = 0$.

Solution. Here $\omega(f,\, \mathbf{a}) = \lim_{r \to 0+} \Omega(f,\, E \cap B_r(\mathbf{a}))$ with $\mathbf{a} \in E$, as defined in Problem 4.96. Assume that f is continuous at $\mathbf{a}$. Then, for each $\varepsilon > 0$ there is a $\delta > 0$ such that $\|f(\mathbf{x}) - f(\mathbf{a})\| < \varepsilon$ whenever $\|\mathbf{x} - \mathbf{a}\| < \delta$. This means that

$$\Omega(f,\, E \cap B_r(\mathbf{a})) \leq 2\varepsilon \quad \text{whenever } 0 < r < \delta.$$

Hence $\omega(f,\, \mathbf{a}) = 0$. Conversely, if $\omega(f,\, \mathbf{a}) = 0$ then given $\varepsilon > 0$ there is a $\delta > 0$ such that

$$\Omega(f,\, E \cap B_r(\mathbf{a})) \leq \varepsilon \quad \text{whenever } 0 < r < \delta.$$

Hence $\|f(\mathbf{x}) - f(\mathbf{a})\| \leq \Omega(f,\, E \cap B_r(\mathbf{a})) \leq \varepsilon$ whenever $0 < r < \delta$. Therefore f is continuous at $\mathbf{a} \in E$.

CHAPTER 5

DERIVATIVES

5.1 FUNCTIONS OF A REAL VARIABLE

In the following problems, the norm on $\mathbb{R}^n$ is the standard Euclidean norm. Also, A always denotes an open interval in $\mathbb{R}$.

5.1 Define $f : \mathbb{R} \to \mathbb{R}^2$ by $f(t) = (\cos 2\pi t, \sin 2\pi t)$. Find $f^{(n)}(t)$ and show that $\langle f^{(n)}(t), f^{(n+1)}(t)\rangle = 0$ for all $t \in \mathbb{R}$ and $n \in \mathbb{N}$.

Solution. Applying ordinary rules of differentiation to the components, we have the following derivatives for non-negative integers k:

$$\begin{aligned}
f^{(4k+1)}(t) &= (2\pi)^{4k+1}(-\sin 2\pi t, \cos 2\pi t); \\
f^{(4k+2)}(t) &= (2\pi)^{4k+2}(-\cos 2\pi t, -\sin 2\pi t); \\
f^{(4k+3)}(t) &= (2\pi)^{4k+3}(\sin 2\pi t, -\cos 2\pi t); \text{ and} \\
f^{(4k)}(t) &= (2\pi)^{4k}(\cos 2\pi t, \sin 2\pi t).
\end{aligned}$$

Analysis in Vector Spaces.
By M. A. Akcoglu, P. F. A. Bartha and D. M. Ha

In fact, the last of these derivatives is identical to f when $k = 0$. The formal proof of the formulas is by induction. The result on inner products reduces to checking the four cases. For example,

$$\begin{aligned}\langle f^{(4k)}(t), f^{(4k+1)}(t)\rangle &= (2\pi)^M(-\sin 2\pi t\cos 2\pi t + \cos 2\pi t \sin 2\pi t)\\ &= 0.\end{aligned}$$

5.3 Assume that $f : A \to \mathbb{R}^n$ is differentiable and that $\|f(t)\| > 0$ for all $t \in A$. Show that $u(t) = f(t)/\|f(t)\|$ is also differentiable and $\langle u(t), u'(t)\rangle = 0$ for all $t \in A$.

Solution. Since each component of f is differentiable and $h(x) = x^{1/2}$ is differentiable for $x > 0$, it follows by ordinary rules of differentiation that

$$\begin{aligned}\|f(t)\| &= \langle f(t), f(t)\rangle^{1/2}\\ &= \left(\sum f_i(t)^2\right)^{1/2}\end{aligned}$$

is differentiable at t provided $\sum f_i(t)^2 > 0$. Hence, again by ordinary rules, $u(t) = f(t)/\|f(t)\|$ is differentiable for all $t \in A$.

Now by the ordinary quotient rule (applied to each component), we have

$$\begin{aligned}u'(t) &= \frac{\|f(t)\|f'(t) - \frac{1}{2\|f(t)\|}f(t)2\langle f(t), f'(t)\rangle}{\|f(t)\|^2}\\ &= \frac{\|f(t)\|^2 f'(t) - f(t)\langle f(t), f'(t)\rangle}{\|f(t)\|^3}.\end{aligned}$$

Consequently,

$$\begin{aligned}\langle u(t), u'(t)\rangle &= \frac{1}{\|f(t)\|^4}(\|f(t)\|^2\langle f(t), f'(t)\rangle - \langle f(t), f(t)\rangle\langle f(t), f'(t)\rangle)\\ &= 0.\end{aligned}$$

5.5 Let X be a normed space. Given $A \in L(X, X)$ and $t \in \mathbb{R}$, let

$$f(t) = e^{tA} = \lim_n \left(I + \frac{t}{1!}A + \frac{t^2}{2!}A^2 + \cdots + \frac{t^n}{n!}A^n\right)$$

as in Problem 4.48. Show that $f : \mathbb{R} \to L(X, X)$ is differentiable. Find $f'(t)$.

Solution. Let $\|\cdot\|$ be the usual norm on $L(X, X)$. Note that for any t, the real-valued series

$$\|I\| + \|\frac{t}{1!}A\| + \|\frac{t^2}{2!}A^2\| + \cdots + \|\frac{t^n}{n!}A^n\| + \cdots$$

converges (absolutely) because it is just the ordinary real-valued series for $e^{|t|\|A\|}$. Next, note that for any t and h, $e^{(t+h)A} = e^{tA}e^{hA}$. This follows from the absolute convergence of each series by expanding the expression for $e^{(t+h)A}$.

Pursuing the analogy with the ordinary exponential function, we show that f is differentiable and $f'(t) = Ae^{tA}$. We have:

$$\begin{aligned}
\|e^{(t+h)A} - e^{tA} - hAe^{tA}\| &= \|e^{tA}\|\|e^{hA} - (I + hA)\| \\
&= \|e^{tA}\|\|\frac{h^2}{2!}A^2 + \cdots\| \\
&\leq |h|^2\|e^{tA}\|\|A^2/2! + \cdots\| \\
&< |h|\varepsilon,
\end{aligned}$$

provided $|h| < \varepsilon/\|e^{\|(1+|t|)A\|}\|$. This proves the conclusion.

5.7 Let $f,\, g : A \to \mathbb{R}^n$ be two differentiable functions. Then show that

$$F = \langle f,\, g\rangle : A \to \mathbb{R}$$

is also differentiable and $F'(t) = \langle f'(t),\, g(t)\rangle + \langle f(t),\, g'(t)\rangle$.

Solution. If $f(t) = \sum f_i(t)\mathbf{e}_i$ and $g(t) = \sum g_i(t)\mathbf{e}_i$, we know that the component functions f_i and g_i are differentiable. So

$$F = \langle f,\, g\rangle = \sum f_i g_i$$

is differentiable by the ordinary rules of differentiation. By the ordinary product rule,

$$\begin{aligned}
F' &= \sum(f_i' g_i + f_i g_i') \\
&= \sum f_i' g_i + \sum f_i g_i' \\
&= \langle f',\, g\rangle + \langle f,\, g'\rangle.
\end{aligned}$$

5.9 Frenet formulas. (See text for definitions.) Show that $(\mathbf{u}(t),\, \mathbf{n}(t),\, \mathbf{b}(t))$ is an orthonormal basis for $\mathbb{R}^3$ and that

$$\begin{aligned}
\mathbf{u}'(t)\, p(t) &= \rho(t)\, \mathbf{n}(t), \\
\mathbf{n}'(t)\, p(t) &= -\rho(t)\, \mathbf{u}(t) - \tau(t)\, \mathbf{b}(t), \text{ and} \\
\mathbf{b}'(t)\, p(t) &= \tau(t)\, \mathbf{n}(t).
\end{aligned}$$

Solution. By problem 5.3, $\mathbf{n}(t)$ is orthogonal to $\mathbf{u}(t)$. The cross product of two vectors in $\mathbb{R}^3$ is orthogonal to those two vectors. Since each of the three vectors is a unit vector, $(\mathbf{u}(t),\, \mathbf{n}(t),\, \mathbf{b}(t))$ is an orthonormal basis for $\mathbb{R}^3$.

From the definition of $\mathbf{n}(t)$, we obtain the first Frenet formula:

$$\rho(t)\,\mathbf{n}(t) = p(t)\|\mathbf{u}'(t)\|\mathbf{n}(t) = p(t)\mathbf{u}'(t).$$

For the second and third formulas, use the fact (problem 5.3) that $\langle \mathbf{n}'(t),\, \mathbf{n}(t)\rangle = 0$ and the fact that $\mathbf{u}(t)$, $\mathbf{n}(t)$, $\mathbf{b}(t)$ form an orthonormal basis to write

$$\mathbf{n}'(t) = \alpha(t)\mathbf{u}(t) + \beta(t)\mathbf{b}(t).$$

Then $\alpha(t) = \langle \mathbf{n}'(t),\, \mathbf{u}(t)\rangle = -\langle \mathbf{n}(t),\, \mathbf{u}'(t)\rangle = -\|\mathbf{u}'(t)\|$, where the second equality follows by differentiating $\langle \mathbf{n}(t),\, \mathbf{u}(t)\rangle = 0$. Similarly, $\beta(t) = -\langle \mathbf{n}(t),\, \mathbf{b}'(t)\rangle$. But

$$\begin{aligned}\mathbf{b}'(t) &= \mathbf{u}'(t) \times \mathbf{n}(t) + \mathbf{u}(t) \times \mathbf{n}'(t)\\ &= \mathbf{u}(t) \times [\alpha(t)\mathbf{u}(t) + \beta(t)\mathbf{b}(t)]\\ &= \mathbf{u}(t) \times \beta(t)\mathbf{b}(t),\end{aligned}$$

which shows that $\mathbf{b}'(t) = -\beta(t)\mathbf{n}(t)$. This gives $\beta(t) = \|\mathbf{b}'(t)\|$. Therefore, we have

$$\begin{aligned}\mathbf{n}'(t)p(t) &= -\|\mathbf{u}'(t)\|p(t)\mathbf{u}(t) - \|\mathbf{b}'(t)\|p(t)\mathbf{b}(t)\\ &= -\rho(t)\mathbf{u}(t) - \tau(t)\mathbf{b}(t),\end{aligned}$$

the second Frenet formula. We also have the third Frenet formula:

$$\mathbf{b}'(t)p(t) = -\|\mathbf{b}'(t)\|p(t)\mathbf{n}(t) = \tau(t)\mathbf{n}(t).$$

5.11 Let X be a Euclidean space and $S : A \to L(X,\, X)$ a differentiable function. Show that for each $\mathbf{a} \in X$ the function $\mathbf{r}(t) = S(t)\,\mathbf{a}$, $t \in A$, is also differentiable and $\mathbf{r}'(t) = S'(t)\,\mathbf{a}$.

Solution. We have

$$\|S(t+h)\mathbf{a} - S(t)\mathbf{a} - hS'(t)\mathbf{a}\| \le \|\mathbf{a}\|\|S(t+h) - S(t) - hS'(t)\|,$$

where the second norm is the usual norm on $L(X,\, X)$. Since S is differentiable, given $\varepsilon > 0$ we can find $\delta > 0$ such that if $|h| < \delta$, then

$$\|S(t+h) - S(t) - hS'(t)\| < |h|\varepsilon/\|\mathbf{a}\|.$$

Then $|h| < \delta$ implies

$$\|\mathbf{r}(t+h) - \mathbf{r}(t) - hS'(t)\mathbf{a}\| < |h|\varepsilon.$$

Hence $\mathbf{r}'(t) = S'(t)\,\mathbf{a}$.

5.13 Let X be a Euclidean space. Let $S : A \to L(X,\, X)$ be such that $\|S(t)\mathbf{v}\| = \|\mathbf{v}\|$ for all $t \in A$ and $\mathbf{v} \in X$. If $S'(t)$ exists, then show that $\langle S'(t)\mathbf{v},\, S(t)\mathbf{v}\rangle = 0$ for all $t \in A$ and $\mathbf{v} \in X$.

Solution. Fix any $\mathbf{v} \in X$ and put $f(t) = \langle S(t)\mathbf{v},\, S(t)\mathbf{v}\rangle = \|S(t)\mathbf{v}\|^2$. The assumption implies that $f(t) = \|\mathbf{v}\|^2$ is constant for all t, so that $f'(t) = 0$ for all $t \in A$. But then by problem 5.7, we have $\langle S'(t)\mathbf{v},\, S(t)\mathbf{v}\rangle = 0$.

5.15 Let $S : A \to L(\mathbb{R}^3,\, \mathbb{R}^3)$ be a differentiable function such that $\|S(t)\mathbf{x}\| = \|\mathbf{x}\|$ for all $t \in A$ and $\mathbf{x} \in \mathbb{R}^3$. Show that for each $t \in A$ there is a unique vector $\mathbf{m}(t) \in \mathbb{R}^3$ such that $S'(t)\mathbf{x} = \mathbf{m}(t) \times S(t)\mathbf{x}$ for all $t \in A$ and for all $\mathbf{x} \in \mathbb{R}^3$.

Solution. 1. Let $Q : \mathbb{R}^3 \to \mathbb{R}^3$ be a linear transformation such that $\mathbf{x} \perp Q\mathbf{x}$ for all $\mathbf{x} \in \mathbb{R}^3$. In this case note that

$$\langle Q(\mathbf{x}+\mathbf{y}),\, (\mathbf{x}+\mathbf{y})\rangle = \langle Q\mathbf{x},\, \mathbf{y}\rangle + \langle \mathbf{x},\, Q\mathbf{y}\rangle = 0. \tag{5.1}$$

We claim that there is an $\mathbf{m} \in \mathbb{R}^3$ such that $Q\mathbf{x} = \mathbf{m} \times \mathbf{x}$ for all $\mathbf{x} \in \mathbb{R}^3$. To see this first assume that there is a unit vector $\mathbf{w}$ such that $Q\mathbf{w} = \mathbf{0}$. Then $Q\mathbf{x} \perp \mathbf{w}$ for all $\mathbf{x} \in \mathbb{R}^3$, since $\langle Q\mathbf{x},\, \mathbf{w}\rangle = -\langle \mathbf{x},\, Q\mathbf{w}\rangle = 0$. Complete $\mathbf{w}$ to a positive orthonormal basis $(\mathbf{u},\, \mathbf{v},\, \mathbf{w})$ in the standard orientation of $\mathbb{R}^3$. Then we see that $Q\mathbf{u} = \omega\mathbf{v}$, since $Q\mathbf{u}$ is orthogonal to both $\mathbf{u}$ and $\mathbf{w}$. Similarly, $Q\mathbf{v} = \beta\mathbf{u}$. Then $\beta = -\omega$, since

$$\langle Q(r\mathbf{u} + s\mathbf{v}),\, (r\mathbf{u} + s\mathbf{v})\rangle = (\omega + \beta)rs = 0 \quad \text{for all} \quad r,\, s \in \mathbb{R}.$$

This shows that $Q\mathbf{x} = \omega\mathbf{w} \times \mathbf{x}$ for all $\mathbf{x} \in \mathbb{R}^3$. Hence let $\mathbf{m} = \omega\mathbf{w}$.

2. To obtain $\mathbf{w}$ let $\mathbf{p}$ be any unit vector. If $Q\mathbf{p} = \mathbf{0}$ then take $\mathbf{w} = \mathbf{p}$. If $\mathbf{q} = Q\mathbf{p} \neq \mathbf{0}$ then let $\mathbf{r}$ be a nonzero vector orthogonal to both $\mathbf{p}$ and $\mathbf{q}$. Then $\langle Q\mathbf{r},\, \mathbf{r}\rangle = 0$ by the hypothesis on Q. Also,

$$\langle Q\mathbf{r},\, \mathbf{p}\rangle = -\langle \mathbf{r},\, Q\mathbf{p}\rangle = -\langle \mathbf{r},\, \mathbf{q}\rangle = 0$$

by the choice of $\mathbf{r}$. Therefore $Q\mathbf{r} = \gamma\mathbf{q}$ for some $\gamma \in \mathbb{R}$. Then we see that $\gamma\mathbf{p} - \mathbf{r}$ is a nonzero vector and $Q(\gamma\mathbf{p} - \mathbf{r}) = \mathbf{0}$. Hence let $\mathbf{w} = (\gamma\mathbf{u} - \mathbf{r})/\|\gamma\mathbf{u} - \mathbf{r}\|$.

3. For the solution of the problem let $t \in A$ be fixed and let $Q = S'(t)S(t)^{-1}$. Then, $\langle QS(t)\mathbf{x}),\, S(t)\mathbf{x}\rangle = \langle S'(t)\mathbf{x},\, S(t)\mathbf{x}\rangle = 0$ by Problem 5.13. But any vector $\mathbf{y} \in \mathbb{R}^3$ can be expressed as $S(t)\mathbf{x}$, since $S(t)$ is an isomorphism. Therefore, by the first part, there is a vector $\mathbf{m}$ such that $Q\mathbf{y} = \mathbf{m} \times \mathbf{y}$ for all $\mathbf{y} \in \mathbb{R}^3$, in particular for all $\mathbf{y} = S(t)\mathbf{x}$. For the uniqueness of $\mathbf{m}$, if $Q\mathbf{y} = \mathbf{m}' \times \mathbf{y}$ for all $\mathbf{y}$ then $(\mathbf{m} - \mathbf{m}') \times \mathbf{y} = \mathbf{0}$ for all $\mathbf{y}$. We see that this implies $\mathbf{m} = \mathbf{m}'$.

5.17 Helicoidal motions. Let $R(t) : \mathbb{R}^3 \to \mathbb{R}^3$ be a rotation given in terms of $\mathbf{w} \in \mathbb{R}^3$ and $\omega \in \mathbb{R}$ as in Problem 5.14. Let $T(t) : \mathbb{R}^3 \to \mathbb{R}^3$ be a translation given by $T(t)\mathbf{x} = \mathbf{x} + t\mathbf{a}$, $\mathbf{x} \in \mathbb{R}^3$, $t \in \mathbb{R}$, where $\mathbf{a} \in \mathbb{R}^3$ is fixed. Then

$$H(t)\mathbf{x} = t\mathbf{c} + R(t)\mathbf{x}, \quad t \in \mathbb{R}, \quad \mathbf{x} \in \mathbb{R}^3$$

is called a *helicoidal motion*. Show that helicoidal motions are rigid motions. If $\mathbf{r}(t) = H(t)\mathbf{x}$, then show that the velocities are given as $\mathbf{r}'(t) = \mathbf{c} + \omega\, \mathbf{w} \times R(t)\mathbf{x}$.

Solution. First:

$$\begin{aligned} \|H(t)\mathbf{a} - H(t)\mathbf{b}\| &= \|(t\mathbf{c} + R(t)\mathbf{a}) - (t\mathbf{c} + R(t)\mathbf{b})\| \\ &= \|R(t)\mathbf{a} - R(t)\mathbf{b}\| \\ &= \|\mathbf{a} - \mathbf{b}\|, \end{aligned}$$

since $R(t)$ is a rigid motion. If $\mathbf{r}(t) = H(t)\mathbf{x}$, then $\mathbf{r}'(t) = H'(t)\mathbf{x} = \mathbf{c} + R'(t)\mathbf{x} = \mathbf{c} + \omega\, \mathbf{w} \times R(t)\mathbf{x}$, by Problem 5.14.

5.19 Instantaneous translations and rotations of Frenet vectors. The Frenet vectors $(\mathbf{u}(t), \mathbf{n}(t), \mathbf{b}(t))$ of a curve $\mathbf{r} : A \to \mathbb{R}^3$ form an orthonormal basis for $\mathbb{R}^3$ for all $t \in A$. Hence, they define a rigid motion as follows. For convenience, assume that $0 \in A$ and define

$$S(t)(\alpha\mathbf{u}(0) + \beta\mathbf{n}(0) + \gamma\mathbf{b}(0)) = \alpha\mathbf{u}(t) + \beta\mathbf{n}(t) + \gamma\mathbf{b}(t)$$

for all $\mathbf{x} = (\alpha, \beta, \gamma) \in \mathbb{R}^3$. Then $M(t)\mathbf{x} = \mathbf{r}(t) + S(t)\mathbf{x}$ defines a rigid motion of $\mathbb{R}^3$. Find the instantaneous translations and rotations of this rigid motion.

Solution. We see that $S(t)$ transforms the orthonormal basis $(\mathbf{u}(0), \mathbf{n}(0), \mathbf{b}(0))$ to another orthonormal basis $(\mathbf{u}(t), \mathbf{n}(t), \mathbf{b}(t))$. Therefore $S(t)$ is an isomorphism at every $t \in A$. Hence, by Problem 5.15, there is a vector $\mathbf{m}(t)$ such that

$$S'(t)\mathbf{x} = \mathbf{m}(t) \times \mathbf{x} \quad \text{for all} \quad \mathbf{x} \in \mathbb{R}^3.$$

But $S'(t)\mathbf{u}(t)$, $S'(t)\mathbf{n}(t)$, and $S'(t)\mathbf{b}(t)$ are given by the Frenet formulas, as obtained in Problem 5.9. Then an easy computation shows that

$$\mathbf{m}(t)p(t) = -\tau(t)\mathbf{u} + \rho(t)\mathbf{b} \quad \text{for all} \quad t \in A.$$

Here, as defined in Problem 5.9, $p(t) = 1/\|\mathbf{r}'(t)\|$ and $\rho(t)$ and $\tau(t)$ are, respectively, the curvature and the torsion of the curve at $\mathbf{r}(t)$.

Remarks. If $M(t)\mathbf{x} = \mathbf{s}(t) + S(t)\mathbf{x}$, $\mathbf{x} \in X$, is a rigid motion of $\mathbb{R}^3$ then the velocities are given as $\mathbf{x}'(t) = \mathbf{s}'(t) + \mathbf{m}(t) \times \mathbf{x}(t)$. Here the origin $\mathbf{0} \in \mathbb{R}^3$ is chosen as the reference point. The trajectory of the origin is $\mathbf{s}(t)$ and its velocities are $\mathbf{s}'(t)$. If one chooses another point $\mathbf{a} \in \mathbb{R}^3$, with the trajectory $\mathbf{a}(t) = M(t)\mathbf{a}$ and the velocities $\mathbf{a}'(t)$ then $\mathbf{x}'(t) = \mathbf{a}'(t) + \mathbf{m}(t) \times (\mathbf{x}(t) - \mathbf{a}(t))$. We see easily that at every instant $t \in A$ there are points with velocities parallel to $\mathbf{m}(t)$. One may consider these points as the points with zero rotational velocities. If $\mathbf{m}(t) \neq \mathbf{0}$ then all such points are on a unique line parallel to $\mathbf{m}(t)$. This is called the *instantaneous* *axis of rotation.* A small computation shows that in the case of the motion of the Frenet vectors this is the line that passes through the point

$$\mathbf{r}(t) + \frac{\rho(t)}{\rho(t)^2 + \tau(t)^2}\mathbf{n}(t)$$

and is parallel to the vector $\mathbf{m}(t)p(t) = -\tau(t)\mathbf{u}(t) + \rho(t)\mathbf{b}(t)$.

Problems on Plane Curves

Let $\mathbf{r} : A \to \mathbb{R}^2 \subset \mathbb{R}^3$ be a plane curve C with the corresponding Frenet vectors $(\mathbf{u}(t), \mathbf{n}(t), \mathbf{b}(t))$. Let $\mathbf{k}$ be a unit vector orthogonal to the subspace $\mathbb{R}^2$ in the vector space $\mathbb{R}^3$.

5.21 The point $\mathbf{e}(t) = \mathbf{r}(t) + (1/\rho(t))\,\mathbf{n}(t)$ is called the *center of curvature of C at the point* $\mathbf{r}(t)$. Then $\mathbf{e} : A \to \mathbb{R}^2 \subset \mathbb{R}^3$ defines another curve E. Show that the unit tangent vector of E at $\mathbf{e}(t)$ is $\pm\mathbf{n}(t)$.

Solution. We see easily that $\tau(t) = 0$ for plane curves. This was also stated in Problem 5.20. Hence $\mathbf{n}'(t)p(t) = -\rho(t)\mathbf{u}$ by Frenet formulas, Problem 5.9. Then

$$\begin{aligned}
\mathbf{e}'(t) &= \mathbf{r}'(t) + \frac{1}{\rho(t)}\mathbf{n}'(t) - \frac{\rho'(t)}{\rho(t)^2}\mathbf{n}(t) \\
&= \frac{1}{p(t)}\mathbf{u} + \frac{1}{\rho(t)}\frac{-\rho(t)}{p(t)}\mathbf{u} - \frac{\rho'(t)}{\rho(t)^2}\mathbf{n}(t) \\
&= -\frac{\rho'(t)}{\rho(t)^2}\mathbf{n}(t).
\end{aligned}$$

Hence the unit tangent vector $\mathbf{e}'(t)/\|\mathbf{e}'(t)\|$ is $\pm\mathbf{n}(t)$.

5.23 Let L and L' be the lines in $\mathbb{R}^2$ passing through the points $\mathbf{r}(t)$ and $\mathbf{r}(t')$ and in the directions of $\mathbf{n}(t)$ and $\mathbf{n}(t')$, respectively. Show that the intersection points $P(t, t')$ of these lines converge to $\mathbf{e}(t)$ as $t \to t'$.

Solution. The points on the lines L and L' are expressed, respectively, as

$$\mathbf{r}(t) + \lambda\mathbf{n}(t) \text{ and } \mathbf{r}(t') + \lambda'\mathbf{n}(t')$$

with scalars λ and λ'. At the point of intersection we have

$$\begin{aligned}
\mathbf{r}(t) + \lambda\mathbf{n}(t) &= \mathbf{r}(t') + \lambda'\mathbf{n}(t') \text{ or} \\
\mathbf{r}(t') - \mathbf{r}(t) &= -\lambda'(\mathbf{n}(t') - \mathbf{n}(t)) + (\lambda - \lambda')\mathbf{n}(t).
\end{aligned}$$

Taking the inner product with $\mathbf{u}(t)$ and dividing by $(t' - t)$ we obtain

$$\left\langle \frac{\mathbf{r}(t') - \mathbf{r}(t)}{t' - t}, \mathbf{u}(t) \right\rangle = -\lambda' \left\langle \frac{\mathbf{n}(t') - \mathbf{n}(t)}{t' - t}, \mathbf{u}(t) \right\rangle. \tag{5.2}$$

Let $t' \to t$. The first ratio in (5.2) converges to $\mathbf{u}(t)/p(t)$. The second ratio converges to $-\rho(t)/p(t)$, as in the solution of problem 5.23. Hence we see that λ' converges to $1/\rho(t)$. Therefore the point of intersection converges to $\mathbf{e}(t) = \mathbf{r}(t) + (1/\rho(t))\mathbf{n}(t)$.

5.2 DIFFERENTIABLE FUNCTIONS

5.25 Consider the xy-plane as a horizontal plane with the z-axis pointing directly upward. Assume that $z = f(x, y) = 10 - x^2 - 2y^2$ describes the surface of a hill.

1. Find the directional derivatives of $z = f(x, y)$ at the point $(x = 2, y = 1)$ and along an arbitrary vector (u, v).
2. Take (u, v) as a unit vector, $u^2+v^2 = 1$. Explain why the directional derivative $f'((2, 1); (u, v))$ can be considered as the slope a hiker would experience if she starts at the point $(2, 1, 4)$ on the hill and moves in the (u, v) direction determined by the horizontal coordinates.
3. Show that there is a plane passing through $(2, 1, 4)$ such that all the slopes on this plane are the same as the corresponding slopes on on the hill.
4. Find the equation of this plane.
5. What are the directions (u, v) of the steepest ascent, the steepest descent, and of zero slope?

Solution. [1.] Working directly from Definition 5.2.11, we have

$$\begin{aligned} f'((2, 1); (u, v)) &= \lim_{t\to 0} \frac{f(2+tu, 1+tv) - f(2, 1)}{t} \\ &= \lim_{t\to 0} \frac{-4tu - t^2u^2 - 4tv - 2t^2v^2}{t} \\ &= -4u - 4v. \end{aligned}$$

[2.] The hiker would move along the curve $g(t) = f((2, 1) + t(u, v))$ whose slope at $t = 0$ is exactly the given directional derivative, by reasoning similar to Lemma 5.2.18.

[3,4.] The claim is that for each unit vector (u, v), the slope along the hill in the (u, v) direction is the same as the slope along the plane in the (u, v) direction. By part [1], we need to find a plane whose slope in the (u, v) direction is $-4u - 4v$. The plane is given as $\{ (2, 1, 4) + s(1, 0, -4) + t(0, 1, -4) \mid s, t \in \mathbb{R} \}$.

[5.] For steepest ascent: maximize $-4u - 4v$ over unit vectors (u, v). Substituting in for v using $u^2 + v^2 = 1$ and applying the first derivative test, we get a maximum in the direction $(u, v) = (-1/\sqrt{2}, -1/\sqrt{2})$. Similarly, for steepest descent, we obtain a minimum for $-4u - 4v$ in the direction $(u, v) = (1/\sqrt{2}, 1/\sqrt{2})$. For zero slope, we have $(u, v) = (-1/\sqrt{2}, 1/\sqrt{2})$ and $(u, v) = (1/\sqrt{2}, -1/\sqrt{2})$.

5.27 Find the directional derivative of

$$f(x,y) = \begin{cases} \dfrac{x^2y^2}{x^4+y^4}, & (x,y) \neq (0,0) \\ 0, & (x,y) = (0,0) \end{cases}$$

at an arbitrary point and in an arbitrary direction, if it exists.

Solution. Consider first the situation at the origin, $(0, 0)$. Here, directional derivatives exist only along the coordinate axes. To see this, note that if either $x = 0$ or $y = 0$, then $f(x, y) = 0$ and the corresponding directional derivatives will be 0. For any other direction, substitute $y = kx$ to obtain $f(x,y) = \frac{k^2x^4}{x^4(k^4+1)} = \frac{k^2}{k^4+1}$. We see that f is constant and non-zero along the line $y = kx$, and therefore discontinuous at $(0, 0)$; hence, the directional derivative does not exist at $(0, 0)$ along any of these lines.

Next consider any point (a, b) that is not the origin. In this case, every directional derivative exists. This will be an easy consequence of results to be proven in section 5.3, but for now the easiest method is to substitute $x = a$ or $y = b$ for directional derivatives parallel to the coordinate axes, or to substitute $y = k(x - a) + b$ for other directional derivatives. This converts $f(x, y)$ into a one-place function that is easily seen to be differentiable (and easily differentiated).

5.29 Same as Problem 5.27 for

$$f(x,y) = \begin{cases} xy \sin \dfrac{1}{x^2+y^2}, & (x,y) \neq (0,0), \\ 0, & (x,y) = (0,0). \end{cases}$$

Solution. First consider the situation at $(0, 0)$. Note that f is identically 0 along the coordinate axes, so that the corresponding directional derivatives will be 0. Along any other line $y = kx$, substitution gives a directional derivative of $\lim_{x\to 0}(kx^2 \sin \frac{1}{x^2(1+k^2)})/x = 0$. So every directional derivative exists and is 0.

Near any other point (a, b), restriction to a straight line via substitution (as in the preceding solution) leads to a one-place function that is easily seen to be differentiable and easily differentiated.

5.31 Let $f : \mathbb{R}^2 \to \mathbb{R}^2$ be the polar coordinates function

$$f(r, \theta) = (x, y) = (r\cos\theta, r\sin\theta).$$

Find the directional derivative of f at the point $(r = 1,\, \theta = \alpha)$ along an arbitrary vector $(a,\, \lambda)$ in the $r\theta$-plane. The result will be a vector in the xy-plane. In particular, find the directional derivatives along the vectors $(1,\, 0)$ and $(0,\, 1)$. Show that these two directional derivatives are orthogonal to each other.

Solution. The required directional derivative is the same as the derivative of

$$g(t) = ((1 + ta)\cos(\alpha + t\lambda), (1 + ta)\sin(\alpha + t\lambda))$$

at $t = 0$. By ordinary differentiation as in section 5.1, this derivative at t is

$$g'(t) = (a\cos(\alpha{+}t\lambda){-}\lambda(1{+}ta)\sin(\alpha{+}t\lambda),\ a\sin(\alpha{+}t\lambda){+}\lambda(1{+}ta)\cos(\alpha{+}t\lambda)),$$

so that the derivative at $t = 0$ is

$$g'(0) = (a\cos\alpha - \lambda\sin\alpha,\ a\sin\alpha + \lambda\cos\alpha).$$

In particular, for the directional derivative along $(1,\, 0)$, we have $a = 1$ and $\lambda = 0$, while for the directional derivative along $(0,\, 1)$, we have $a = 0$ and $\lambda = 1$. These two directional derivatives are therefore $(\cos\alpha,\, \sin\alpha)$ and $(-\sin\alpha,\, \cos\alpha)$. Since their inner product is 0, they are orthogonal.

5.33 Let $f(\rho,\, \varphi,\, \theta) = (x,\, y,\, z)$ be the spherical coordinates function

$$x = \rho\sin\varphi\cos\theta,\quad y = \rho\sin\varphi\sin\theta,\quad z = \rho\cos\varphi.$$

Find the directional derivative of f at the point $(\rho = 1,\, \theta = \alpha,\, \varphi = \beta)$ along an arbitrary vector $(a,\, \lambda,\, \mu)$ in the $r\theta\varphi$-space. In particular, find the directional derivatives along the vectors $(1,\, 0,\, 0)$, $(0,\, 1,\, 0)$, and $(0,\, 0,\, 1)$. Show that these three directional derivatives are orthogonal to each other.

Solution. Let $(\mathbf{i},\, \mathbf{j},\, \mathbf{k})$ be the standard basis for the xyz-space. Let

$$\mathbf{r}(\rho,\, \theta,\, \varphi) = \rho\sin\varphi\cos\theta\,\mathbf{i} + \rho\sin\varphi\sin\theta\,\mathbf{j} + \rho\cos\varphi\,\mathbf{k}.$$

The required directional derivative is

$$\lim_{t\to 0}\frac{1}{t}\left(\mathbf{r}(1 + at,\, \alpha + t\lambda,\, \beta + t\mu) - \mathbf{r}(1,\, \alpha,\, \beta)\right). \tag{5.3}$$

This is computed easily by the ordinary rules of differentiation. In particular the directional derivatives at $(1,\, \alpha,\, \beta)$ in the directions of $(1,\, 0,\, 0)$, $(0,\, 1,\, 0)$, and $(0,\, 0,\, 1)$ are ,respectively

$$\begin{aligned}
\mathbf{e}_1(1,\, \alpha,\, \beta) &= \sin\beta\cos\alpha\,\mathbf{i} + \sin\beta\sin\alpha\,\mathbf{j} + \cos\beta\,\mathbf{k},\\
\mathbf{e}_2(1,\, \alpha,\, \beta) &= -\sin\beta\sin\alpha\,\mathbf{i} + \sin\beta\cos\alpha\,\mathbf{j}, \text{ and}\\
\mathbf{e}_3(1,\, \alpha,\, \beta) &= \cos\beta\cos\alpha\,\mathbf{i} + \cos\beta\sin\alpha\,\mathbf{j} - \sin\beta\,\mathbf{k}.
\end{aligned}$$

An easy verification shows that these vectors form an orthogonal basis for the xyz-space. In terms of this basis the limit in (5.3) is obtained as

$$a\mathbf{e}_1(1,\, \alpha,\, \beta) + \lambda\mathbf{e}_2(1,\, \alpha,\, \beta) + \mu\mathbf{e}_3(1,\, \alpha,\, \beta).$$

5.3 EXISTENCE OF DERIVATIVES

5.35 Let $f : \mathbb{R}^3 \to \mathbb{R}$ be defined as $f(x, y, z) = 1$ if $x = t$, $y = t^2$, and $z = t^3$ for some $t > 0$, and $f(x, y, z) = 0$ otherwise. Show that f has a restricted derivative at the origin along any one- or two-dimensional subspace. Show that f is not differentiable at the origin.

Solution. Plainly, f is not continuous at the origin and therefore it cannot be differentiable at the origin. For the main part of the question, it suffices to show that for any one- or two-dimensional subspace U containing the origin, there will be a neighborhood of the origin that does not meet any point of the form (t, t^2, t^3) for positive t, and hence a neighborhood on which the restricted function f_U is identically 0. It then follows that f_U is differentiable at the origin, with derivative 0.

Note first that if $\mathbf{u} = (a, b, c)$ is any unit vector in $\mathbb{R}^3$, then there is at most one $\alpha \neq 0$ such that $\alpha\mathbf{u}$ is of the form (t, t^2, t^3), namely, $\alpha = b/a^2$. The claim about neighborhoods follows at once for one-dimensional subspaces: on any line through the origin, there will be a neighborhood of the origin that does not contain points of the form (t, t^2, t^3) with $t > 0$.

Any plane containing the origin is determined as the set of points satisfying $Ax + By + Cz = 0$ for constants A, B, and C. Then $At + Bt^2 + Ct^3 = 0$ only if $t = 0$ or $Ct^2 + Bt + A = 0$. Hence there are at most two non-trivial points on the plane of the form (t, t^2, t^3). The claim about neighborhoods follows for two-dimensional subspaces.

5.37 Let $f(x, y) = 1$ if $y < x^2 < 2y$ and $f(x, y) = 0$ otherwise. Show that f has directional derivatives at the origin in any direction. Is f differentiable at the origin?

Solution. $f(x, y) = 1$ on the region between the parabolas $y = x/2$ and $y = x$, and 0 elsewhere. Clearly, f is discontinuous at the origin and therefore not differentiable at the origin. Along the x- or y-axis, f is identically 0 and hence the directional derivative is 0. Along the line $y = kx$ with $k > 0$, we have $kx > x^2$ if $x < k$, so that $f(x, kx) = 0$ if $x < k$. This shows that the directional derivative along the line $y = kx$ is also 0, and a similar argument works if $k < 0$. Note that even though all directional derivatives exist and are identical to 0, the full derivative of f does not exist.

5.39 Suppose that $f : \mathbb{R}^2 \to \mathbb{R}$ has a directional derivative at the origin along the vector $(1, 0)$. Assume that for all $x \in \mathbb{R}$ the directional derivative of f at the point $(x, 0)$ and in the direction of $(1, 1)$ exists and is equal to $p(x)$. Show that if $p : \mathbb{R} \to \mathbb{R}$ is continuous, then f is differentiable at $(0, 0)$.

Solution. There is a mistake in this problem. The following is a counterexample. Let $f(x, y) = 1$ in the region $\sqrt{|x|} < |y|$ and 0 elsewhere. Then f is discontinuous (and hence not differentiable) at the origin. But f has a directional derivative of 0 along $(1, 0)$ and $p(x)$ is identically 0.

For the correct result, the second assumption should be that for all (x, y) in a neighborhood A of $(0, 0)$, the directional derivative of f at (x, y) in the direction of $(1, 1)$ exists and is equal to $p(x, y)$, and $p : A \to \mathbb{R}$ is continuous. Then we can apply Lemma 5.3.2, with $\mathbf{e} = (1, 1)$, $\mathbf{a} = (0, 0)$, and $h = p$. Hence, given $\varepsilon > 0$, there is a $\delta > 0$ such that

$$\|f((x, y) + t\mathbf{e}) - f(x, y) - tp(0, 0)\| \leq \varepsilon|t|$$

whenever $\|(x, y)\| < \delta$ and $|t| < \delta$.

Write $q(0, 0)$ for the directional derivative at the origin along the vector $(1, 0)$. We shall show that f is differentiable at $(0, 0)$ with derivative

$$S(x, y) = q(0, 0)(x - y) + p(0, 0)y.$$

First we may suppose that δ above is chosen small enough so that we also have

$$\|f(t, 0) - f(0, 0) - tq(0, 0)\| \leq \varepsilon|t|$$

whenever $|t| < \delta$. Second, notice that $(x, y) = (x - y, 0) + y\mathbf{e}$. It follows that if both $|x| < \delta/2$ and $|y| < \delta/2$, then $|x - y| < \delta$ and we have the following:

$$\begin{aligned}
\|f(x, y) - f(0, 0) - S(x, y)\| &= \|f(x, y) - f(0, 0) \\
&\quad -q(0, 0)(x - y) - p(0, 0)y\| \\
&\leq \|f((x - y, 0) + y\mathbf{e}) - f(x - y, 0) \\
&\quad -p(0, 0)y\| \\
&\quad +\|f(x - y, 0) - f(0, 0) - q(0, 0)(x - y)\| \\
&\leq \varepsilon|y| + \varepsilon|x - y| \\
&< 4\varepsilon\|(x, y)\|.
\end{aligned}$$

The final step uses the fact that $(|x| + |y|) \leq 2\|(x, y)\|$.

5.4 PARTIAL DERIVATIVES

5.41 Same as Problem 5.40 for $(u, v) = (xy, y/x)$.

Solution.

$$\nabla u(x, y) = \left(\frac{\partial u}{\partial x}, \frac{\partial u}{\partial y}\right) = (y, x).$$

$$\nabla v(x, y) = \left(\frac{\partial v}{\partial x}, \frac{\partial v}{\partial y}\right) = (\frac{-y}{x^2}, \frac{1}{x}),$$

if $x \neq 0$. The Jacobian matrix at (x, y), where $x \neq 0$, is given by

$$\partial(u, v)/\partial(x, y) = \begin{pmatrix} y & x \\ -y/x^2 & 1/x \end{pmatrix}.$$

5.43 Same as Problem 5.40 for

$$(u, v) = (p(x, y) + q(x, y), p(x, y) - q(x, y)),$$

where $p(x, y) = ((x+1)^2 + y^2)^{1/2}$ and $q(x, y) = ((x-1)^2 + y^2)^{1/2}$.

Solution. By straightforward differentiation:

$$\frac{\partial u}{\partial x} = \frac{x+1}{p(x, y)} + \frac{x-1}{q(x, y)}$$

$$\frac{\partial u}{\partial y} = \frac{y}{p(x, y)} + \frac{y}{q(x, y)}$$

$$\frac{\partial v}{\partial x} = \frac{x+1}{p(x, y)} - \frac{x-1}{q(x, y)}$$

$$\frac{\partial v}{\partial y} = \frac{y}{p(x, y)} - \frac{y}{q(x, y)}.$$

These are defined everywhere except $(-1, 0)$ and $(1, 0)$. The Jacobian matrix at all other points (x, y) is given by

$$\frac{\partial(u, v)}{\partial(x, y)} = \begin{pmatrix} \frac{x+1}{p(x, y)} + \frac{x-1}{q(x, y)} & \frac{y}{p(x, y)} + \frac{y}{q(x, y)} \\ \frac{x+1}{p(x, y)} - \frac{x-1}{q(x, y)} & \frac{y}{p(x, y)} - \frac{y}{q(x, y)} \end{pmatrix}.$$

The gradient $\nabla u(x, y)$ is the first row of the matrix, and the gradient $\nabla v(x, y)$ is the second row.

5.45 A particle moves in the xy-plane according to the law of motion

$$\mathbf{s}(t) = (x(t), y(t)), \text{ where } x(t) = e^t \cos 2t \text{ and } y(t) = e^t \sin 2t.$$

Find the coordinates of the velocity $\mathbf{s}'(t)$ and acceleration $\mathbf{s}''(t)$ of this particle with respect to the basis $\mathbf{e}_1(e^t, 2t)$ and $\mathbf{e}_2(e^t, 2t)$ defined in Problem 5.44.

Solution. From Problem 5.44,

$$\mathbf{e}_1(e^t, 2t) = (\cos 2t, \sin 2t)$$

and

$$\mathbf{e}_2(e^t,\, 2t) = (-e^t \sin 2t,\, e^t \cos 2t).$$

Then

$$\begin{aligned} \mathbf{s}'(t) &= (x'(t),\, y'(t)) \\ &= (e^t \cos 2t - 2e^t \sin 2t,\, e^t \sin 2t + 2e^t \cos 2t) \\ &= e^t \mathbf{e}_1(e^t,\, 2t) + 2\mathbf{e}_2(e^t,\, 2t). \end{aligned}$$

Differentiating again yields

$$\mathbf{s}''(t) = -3e^t \mathbf{e}_1(e^t,\, 2t) + 4\mathbf{e}_2(e^t,\, 2t).$$

As an alternative approach, we can compute the derivatives of the vectors $\mathbf{e}_i$ in terms of each other. More formally, if we put $f_1(t) = \mathbf{e}_1(e^t,\, 2t)$ and $f_2(t) = \mathbf{e}_2(e^t,\, 2t)$, then we can compute

$$\begin{aligned} e^t f_1'(t) &= 2f_2(t) \\ f_2'(t) &= -2e^t f_1(t) + f_2(t). \end{aligned}$$

Since $\mathbf{s}(t) = e^t f_1(t)$, we obtain the stated results for $\mathbf{s}'(t)$ and $\mathbf{s}''(t)$.

5.47 A particle moves in the xyz-space according to the law of motion

$$\mathbf{s}(t) = (x(t),\, y(t),\, z(t)) \text{ where}$$

$$x(t) = e^t \cos 2t, \quad y(t) = e^t \sin 2t, \quad z(t) = 3t.$$

Express the velocity $\mathbf{s}'(t)$ and acceleration $\mathbf{s}''(t)$ of this particle in terms of the basis $\mathbf{e}_1(e^t,\, 2t,\, 3t)$, $\mathbf{e}_2(e^t,\, 2t,\, 3t)$, and $\mathbf{e}_3(e^t,\, 2t,\, 3t)$ defined in Problem 5.46.

Solution. From Problem 5.46,

$$\mathbf{e}_1(e^t,\, 2t,\, 3t) = (\cos 2t,\, \sin 2t,\, 0),$$

$$\mathbf{e}_2(e^t,\, 2t,\, 3t) = (-e^t \sin 2t,\, e^t \cos 2t,\, 0),$$

and

$$\mathbf{e}_3(e^t,\, 2t,\, 3t) = (0,\, 0,\, 1).$$

Then

$$\begin{aligned} \mathbf{s}'(t) &= (x'(t),\, y'(t),\, z'(t)) \\ &= (e^t \cos 2t - 2e^t \sin 2t,\, e^t \sin 2t + 2e^t \cos 2t,\, 3) \\ &= e^t \mathbf{e}_1(e^t,\, 2t,\, 3t) + 2\mathbf{e}_2(e^t,\, 2t,\, 3t) + 3\mathbf{e}_3(e^t,\, 2t,\, 3t). \end{aligned}$$

Differentiating again yields

$$\mathbf{s}''(t) = -3e^t\mathbf{e}_1(e^t,\, 2t,\, 3t) + 4\mathbf{e}_2(e^t,\, 2t,\, 3t).$$

5.49 A particle moves in the xyz-space according to the law of motion

$$\mathbf{s}(t) = (x(t),\, y(t),\, z(t)), \quad \text{where}$$

$$x(t) = e^t \sin 3t \cos 2t, \quad y(t) = e^t \sin 3t \sin 2t, \quad z(t) = 3t \cos 3t.$$

Express the velocity $\mathbf{s}'(t)$ and acceleration $\mathbf{s}''(t)$ of this particle in terms of the basis $\mathbf{e}_1(\rho,\, \varphi,\, \theta)$, $\mathbf{e}_2(\rho,\, \varphi,\, \theta)$, and $\mathbf{e}_3(\rho,\, \varphi,\, \theta)$ defined in Problem 5.48.

Solution. Here we have $\rho = e^t$, $\varphi = 3t$, and $\theta = 2t$. From Problem 5.48,

$$\mathbf{e}_1(e^t,\, 3t,\, 2t) = (\sin 3t \cos 2t,\, \sin 3t \sin 2t,\, \cos 3t),$$

$$\mathbf{e}_2(e^t,\, 3t,\, 2t) = (e^t \cos 3t \cos 2t,\, e^t \cos 3t \sin 2t,\, -e^t \sin 3t),$$

and

$$\mathbf{e}_3(e^t,\, 3t,\, 2t) = (-e^t \sin 3t \sin 2t,\, e^t \sin 3t \cos 2t,\, 0).$$

We can compute the derivatives of the vectors $\mathbf{e}_i$ in terms of each other (provided $\sin 3t \neq 0$, so that the three vectors constitute an orthogonal set). More formally, if we put $f_1(t) = \mathbf{e}_1(e^t,\, 3t,\, 2t)$, $f_2(t) = \mathbf{e}_2(e^t,\, 3t,\, 2t)$, and $f_3(t) = \mathbf{e}_3(e^t,\, 3t,\, 2t)$, then we can compute

$$\begin{aligned}
e^t f_1'(t) &= 3f_2(t) + 2f_3(t) \\
f_2'(t) &= -3e^t f_1(t) + f_2(t) + 2\cot 3t\, f_3(t) \\
f_3'(t) &= -2e^t \sin^2 3t\, f_1(t) - 2\sin 3t \cos 3t\, f_2(t) + (1 + 3\cot 3t) f_3(t).
\end{aligned}$$

Since $\mathbf{s}(t) = e^t\mathbf{e}_1(e^t,\, 3t,\, 2t) = e^t f_1(t)$, we have

$$\begin{aligned}
\mathbf{s}'(t) &= e^t f_1(t) + e^t f_1'(t) \\
&= e^t f_1(t) + 3f_2(t) + 2f_3(t) \\
&= e^t\mathbf{e}_1(e^t,\, 3t,\, 2t) + 3\mathbf{e}_2(e^t,\, 3t,\, 2t) + 2\mathbf{e}_3(e^t,\, 3t,\, 2t).
\end{aligned}$$

Differentiating again yields

$$\begin{aligned}
\mathbf{s}''(t) &= e^t f_1(t) + e^t f_1'(t) + 3f_2'(t) + 2f_3'(t) \\
&= e^t f_1(t) + (3f_2(t) + 2f_3(t)) \\
&\quad +3[f_2(t) - 3e^t f_1(t) + 2\cot 3t\, f_3(t)] \\
&\quad +2[-2e^t \sin^2 3t\, f_1(t) - 2\sin 3t \cos 3t\, f_2(t) + (1 + 3\cot 3t) f_3(t)] \\
&= -4e^t(2 + \sin^2 3t) f_1(t) \\
&\quad +(6 - 4\sin 3t \cos 3t) f_2(t) \\
&\quad +(4 + 12\cot 3t) f_3(t).
\end{aligned}$$

5.5 RULES OF DIFFERENTIATION

5.51 Let $\varphi : X \to \mathbb{R}$ be a differentiable function. If $\varphi(\mathbf{a}) \neq 0$, then show that $\nabla(1/\varphi)(\mathbf{a}) = -(1/\varphi(\mathbf{a})^2)\nabla\varphi(\mathbf{a})$.

Solution. We have $\nabla f(\mathbf{a}) = \sum_i (\partial f/\partial x_i)(\mathbf{a})\mathbf{e}_i$ as observed in Remarks 5.4.12. Here x_is are the coordinates with respect to the standard basis $(\mathbf{e}_1, \ldots, \mathbf{e}_n)$. Then

$$\frac{\partial f}{\partial x_i}(\mathbf{a}) = -\frac{1}{\varphi(\mathbf{a})^2}\frac{\partial \varphi}{\partial x_i}(\mathbf{a})$$

by the chain rule, Theorem 5.5.6. Then the solution follows.

5.53 Let A be an open set in the Euclidean space $\mathbb{R}^n$. Let $f : A \to \mathbb{R}$ be a twice continuously differentiable function. Then the *Laplacian* of f is defined as

$$\Delta f(\mathbf{x}) = \frac{\partial^2 f}{\partial x_1^2}(\mathbf{x}) + \cdots + \frac{\partial^2 f}{\partial x_n^2}(\mathbf{x}),$$

where $\mathbf{x} = (x_1, \ldots, x_n) \in A$. Let $n = 2$ and let $(x_1, x_2) = (x, y)$. Express $f(x, y)$ in terms of polar coordinates $x = r\cos\theta$ and $y = r\sin\theta$ as $f(x, y) = F(r, \theta)$. Show that

$$\Delta f = \frac{\partial^2 f}{\partial x^2} + \frac{\partial^2 f}{\partial y^2} = \frac{\partial^2 F}{\partial r^2} + \frac{1}{r}\frac{\partial F}{\partial r} + \frac{1}{r^2}\frac{\partial^2 F}{\partial \theta^2}.$$

Solution. Note that in the text there was a misprint in the expression of Laplacian in polar coordinates. The correct form is given above. Since f is twice continuously differentiable, all second partial derivatives exist. First we compute the partial derivatives:

$$\begin{aligned}\frac{\partial f}{\partial r} &= \frac{\partial f}{\partial x}\frac{\partial x}{\partial r} + \frac{\partial f}{\partial y}\frac{\partial y}{\partial r}\\ &= \cos\theta\frac{\partial f}{\partial x} + \sin\theta\frac{\partial f}{\partial y}\end{aligned}$$

and

$$\begin{aligned}\frac{\partial f}{\partial \theta} &= \frac{\partial f}{\partial x}\frac{\partial x}{\partial \theta} + \frac{\partial f}{\partial y}\frac{\partial y}{\partial \theta}\\ &= -r\sin\theta\frac{\partial f}{\partial x} + r\cos\theta\frac{\partial f}{\partial y}.\end{aligned}$$

Next,

$$\begin{aligned}
\frac{\partial^2 f}{\partial r^2} &= \cos\theta\left(\frac{\partial^2 f}{\partial x^2}\frac{\partial x}{\partial r} + \frac{\partial^2 f}{\partial y\partial x}\frac{\partial y}{\partial r}\right) \\
&\quad + \sin\theta\left(\frac{\partial^2 f}{\partial x\partial y}\frac{\partial x}{\partial r} + \frac{\partial^2 f}{\partial y^2}\frac{\partial y}{\partial r}\right) \\
&= \cos^2\theta\frac{\partial^2 f}{\partial x^2} + \cos\theta\sin\theta\left(\frac{\partial^2 f}{\partial y\partial x} + \frac{\partial^2 f}{\partial x\partial y}\right) + \sin^2\theta\frac{\partial^2 f}{\partial y^2}.
\end{aligned}$$

Finally, differentiating the expression for $(\partial f/\partial\theta)$, we have

$$\begin{aligned}
\frac{\partial^2 f}{\partial \theta^2} &= -r\sin\theta\left(\frac{\partial^2 f}{\partial x^2}(-r\sin\theta) + \frac{\partial^2 f}{\partial y\partial x}r\cos\theta\right) - r\cos\theta\frac{\partial f}{\partial x} \\
&\quad + r\cos\theta\left(\frac{\partial^2 f}{\partial x\partial y}(-r\sin\theta) + \frac{\partial^2 f}{\partial y^2}r\cos\theta\right) - r\sin\theta\frac{\partial f}{\partial y} \\
&= r^2\sin^2\theta\frac{\partial^2 f}{\partial x^2} - r^2\sin\theta\cos\theta\left(\frac{\partial^2 f}{\partial y\partial x} + \frac{\partial^2 f}{\partial x\partial y}\right) \\
&\quad + r^2\cos^2\theta\frac{\partial^2 f}{\partial y^2} - r\cos\theta\frac{\partial f}{\partial x} - r\sin\theta\frac{\partial f}{\partial y}.
\end{aligned}$$

From these equations, we see that

$$\frac{\partial^2 f}{\partial r^2} + \frac{1}{r}\frac{\partial f}{\partial r} + \frac{1}{r^2}\frac{\partial^2 f}{\partial \theta^2} = \frac{\partial^2 f}{\partial x^2} + \frac{\partial^2 f}{\partial y^2} = \Delta f.$$

5.55 Let $k \in \mathbb{N}$. Let $f : \mathbb{R}^n \to \mathbb{R}$ be a differentiable function. If $f(t\mathbf{x}) = t^k f(\mathbf{x})$ for all $\mathbf{x} \in \mathbb{R}^n$ and for all $t \in \mathbb{R}$, then show that $\langle \nabla f(\mathbf{x}), \mathbf{x}\rangle = kf(\mathbf{x})$.

Solution. There was a misprint here. The hypothesis on f should have been that $f(t\mathbf{x}) = t^k f(\mathbf{x})$ for all $\mathbf{x} \in \mathbb{R}^n$ and for all $t \in \mathbb{R}$. For the solution, let $\mathbf{x} \in \mathbb{R}^n$ be fixed. Define $\varphi : \mathbb{R} \to \mathbb{R}$ by $\varphi(t) = f(t\mathbf{x})$. Then φ is the composition of the functions $\psi : \mathbb{R} \to \mathbb{R}^n$ defined as $\psi(t) = t\mathbf{x}$ and $f : \mathbb{R}^n \to \mathbb{R}$. Hence obtain $\varphi'(t)$ by the chain rule, Theorem 5.5.6, as

$$\varphi'(t) = f'(t\mathbf{x})\psi'(t) = f'(t\mathbf{x})\mathbf{x} = \langle \nabla f(t\mathbf{x}), \mathbf{x}\rangle.$$

But, also, $\varphi(t) = t^k f(\mathbf{x})$ and $\varphi'(t) = kt^{k-1}f(\mathbf{x})$. Therefore

$$\langle \nabla f(t\mathbf{x}), \mathbf{x}\rangle = kt^{k-1}f(\mathbf{x}).$$

The solution follows by letting $t = 1$.

5.57 Let $g : \mathbb{R} \to \mathbb{R}$ be a differentiable function. Define $f : X \times Y \to \mathbb{R}$ as $f(\mathbf{x}, \mathbf{y}) = g(\langle T\mathbf{x}, \mathbf{y}\rangle)$, where $T : X \to Y$ is a linear transformation. Find ∇f.

Solution. Let $U = V = \mathbb{R}^n$. Define $h : U \times V \to \mathbb{R}$ as $h_1(\mathbf{u}, \mathbf{v}) = \langle \mathbf{u}, \mathbf{v} \rangle$. Let $\mathbf{p}_i$s and $\mathbf{q}_i$s be, respectively, the standard bases for $U = \mathbb{R}^n$ and $V = \mathbb{R}^n$. We see that

$$\nabla h_1(\mathbf{a}, \mathbf{b}) = \sum\nolimits_i b_i \mathbf{q}_i + \sum\nolimits_j a_j \mathbf{p}_j = (\mathbf{b}, \mathbf{a}) \in \mathbb{R}^n \times \mathbb{R}^n.$$

Hence we obtain $h_1'(\mathbf{a}, \mathbf{b})(\mathbf{u}, \mathbf{v}) = \langle \mathbf{b}, \mathbf{u} \rangle + \langle \mathbf{a}, \mathbf{v} \rangle$. Now let $T : \mathbb{R}^m \to \mathbb{R}^n$ be a linear transformation. The mapping that takes $(\mathbf{x}, \mathbf{v})$ to $(T\mathbf{x}, \mathbf{v})$ is a linear transformation of $X \times V$ to $U \times V$. Hence if we define

$$h : X \times V \to \mathbb{R} \text{ by } h(\mathbf{x}, \mathbf{v}) = \langle T\mathbf{x}, \mathbf{v} \rangle$$

then the chain rule, Theorem 5.5.6, gives

$$\begin{aligned} h'(\mathbf{a}, \mathbf{b})(\mathbf{x}, \mathbf{y}) &= \langle T\mathbf{x}, \mathbf{b} \rangle + \langle T\mathbf{a}, \mathbf{y} \rangle && (5.4) \\ &= \langle T^*\mathbf{b}, \mathbf{x} \rangle + \langle T\mathbf{a}, \mathbf{y} \rangle. && (5.5) \end{aligned}$$

Here $T^* : \mathbb{R}^n \to \mathbb{R}^m$ is the adjoint of $T : \mathbb{R}^m \to \mathbb{R}^n$, defined by the condition that

$$\langle T\mathbf{x}, \mathbf{u} \rangle_{\mathbb{R}^n} = \langle \mathbf{x}, T^*\mathbf{u} \rangle_{\mathbb{R}^m}$$

for all $\mathbf{x} \in \mathbb{R}^m$ and $\mathbf{u} \in \mathbb{R}^n$. The existence and the uniqueness of T^* was obtained in Theorem 3.6.5. Hence (5.4) and (5.5) show that

$$\nabla h(\mathbf{a}, \mathbf{b}) = (T^*\mathbf{b}, T\mathbf{a}) \in \mathbb{R}^m \times \mathbb{R}^n.$$

We obtain $\nabla f(\mathbf{a}, \mathbf{b})$ by another application of the chain rule, Theorem 5.5.6. Since

$$\begin{aligned} f'(\mathbf{a}, \mathbf{b})(\mathbf{x}, \mathbf{y}) &= g'(\langle T\mathbf{a}, \mathbf{b} \rangle) h'(\mathbf{a}, \mathbf{b})(\mathbf{x}, \mathbf{y}) \\ &= g'(\langle T\mathbf{a}, \mathbf{b} \rangle)(\langle T\mathbf{x}, \mathbf{b} \rangle_{\mathbb{R}^n} + \langle T\mathbf{a}, \mathbf{y} \rangle_{\mathbb{R}^n}) \\ &= g'(\langle T\mathbf{a}, \mathbf{b} \rangle)(\langle T^*\mathbf{b}, \mathbf{x} \rangle_{\mathbb{R}^m} + \langle T\mathbf{a}, \mathbf{y} \rangle_{\mathbb{R}^n}) \\ &= g'(\langle T\mathbf{a}, \mathbf{b} \rangle)\langle (T^*\mathbf{b}, T\mathbf{a}), (\mathbf{x}, \mathbf{y}) \rangle_{\mathbb{R}^m \times \mathbb{R}^n} \end{aligned}$$

we see that $\nabla f(\mathbf{a}, \mathbf{b}) = g'(\langle T\mathbf{a}, \mathbf{b} \rangle)(T^*\mathbf{b}, T\mathbf{a})$.

5.6 DIFFERENTIATION OF PRODUCTS

5.59 Let $T : X \to X$ be a linear transformation. Let $f(\mathbf{x}) = \langle T\mathbf{x}, \mathbf{x} \rangle$, $\mathbf{x} \in X$. Show that $\langle \nabla f(\mathbf{a}), \mathbf{x} \rangle = \langle T\mathbf{a}, \mathbf{x} \rangle + \langle T\mathbf{x}, \mathbf{a} \rangle$, $\mathbf{a}, \mathbf{x} \in X$.

Solution. This follows directly from the solution of Problem 5.57. Take $g(r) = r$, $n = m$, and evaluate $\nabla f(\mathbf{a}, \mathbf{a})$ in that problem to obtain

$$\begin{aligned} \nabla f(\mathbf{a}, \mathbf{a}) &= (T^*\mathbf{a}, T\mathbf{a}) \text{ and, therefore,} \\ f'(\mathbf{a}, \mathbf{a})(\mathbf{x}, \mathbf{x}) &= \langle T^*\mathbf{a}, \mathbf{x} \rangle + \langle T\mathbf{a}, \mathbf{x} \rangle \\ &= \langle \mathbf{a}, T\mathbf{x} \rangle + \langle T\mathbf{a}, \mathbf{x} \rangle. \end{aligned}$$

In the present problem $f(\mathbf{x})$ stands for $f(\mathbf{x}, \mathbf{x})$ in Problem 5.57. Hence the solution follows.

5.61 Let $T \in L(X, Y)$ and let $\varphi : X \to \mathbb{R}$ be a differentiable function. Find ∇f for $f(\mathbf{x}) = \langle \varphi(\mathbf{x})T\mathbf{x}, \mathbf{x}\rangle$, $\mathbf{x} \in X$.

Solution. We have $f(\mathbf{x}) = \langle \varphi(\mathbf{x})T\mathbf{x}, \mathbf{x}\rangle = \varphi(\mathbf{x})\langle T\mathbf{x}, \mathbf{x}\rangle$. This is the product of two real valued functions φ and $\psi(\mathbf{x}) = \langle T\mathbf{x}, \mathbf{x}\rangle$. Hence, by the general product rule, Theorem 5.6.4,

$$f'(\mathbf{a}) = \varphi(\mathbf{a})\psi'(\mathbf{a}) + \psi(\mathbf{a})\varphi'(\mathbf{a}).$$

Then, by the result of Problem 5.59,

$$\nabla f(\mathbf{a}) = \varphi(\mathbf{a})T\mathbf{a} + \varphi(\mathbf{a})T^*\mathbf{a} + \langle T\mathbf{a}, \mathbf{a}\rangle \nabla\varphi(\mathbf{a}).$$

CHAPTER 6

DIFFEOMORPHISMS AND MANIFOLDS

6.1 THE INVERSE FUNCTION THEOREM

6.1 Let

$$\begin{aligned} A &= \{\mathbf{x} = (x, y) \in \mathbb{R}^2 \mid x > 0, y > 0\}, \\ B &= \{\mathbf{x} = (x, y) \in \mathbb{R}^2 \mid y > 0\}. \end{aligned}$$

Let

$$f(\mathbf{x}) = (x^2 - y^2, 2xy) \quad \text{for all } \mathbf{x} \in A.$$

Show that f is a diffeomorphism from A onto B.

Solution We will first show that f is one-to-one. Let (x, y) and (u, v) be in A and suppose that $f(x, y) = f(u, v)$. If $x < u$, then since $x^2 - y^2 = u^2 - v^2$, we must have $y < v$. This implies that $xy < uv$, a contradiction. Similarly, $x > u$ leads

Analysis in Vector Spaces.
By M. A. Akcoglu, P. F. A. Bartha and D. M. Ha

to a contradiction. Hence, we must have $x = u$. But since $xy = uv$, we have $y = v$ also. Thus, f is one-to-one. To show that $f(A) = B$, let $\mathbf{p}$ be in B be arbitrary. Then $\mathbf{p} = r(\cos\theta, \sin\theta)$ for some $0 < \theta < \pi$ and some $r > 0$. Let $x = \sqrt{r}\cos\theta/2, y = \sqrt{r}\sin\theta/2$. Then $(x, y) \in A$ and

$$\begin{aligned} x^2 - y^2 &= r(\cos^2\theta/2 - \sin^2\theta/2) = r\cos\theta \\ 2xy &= r(2\cos\theta/2\sin\theta/2) = r\sin\theta. \end{aligned}$$

Hence, $f(x, y) = (r\cos\theta, r\sin\theta) = \mathbf{p}$. This shows that $f : A \to B$ has an inverse $g : B \to A$. Finally, let $(a, b) \in A$. Then

$$f'(a,\, b)(x,\, y) = (2ax - 2by,\, 2bx + 2ay)$$

shows that $f'(a, b) : X \to Y$ is invertible. Also, $f' : A \to L(X, Y)$ is continuous since the components of f are polynomials. Hence, by the inverse function theorem, Theorem 6.1.4, the inverse function is continuously differentiable. Therefore f is a diffeomorphism $A \to f(A) = B$.

6.3 Let D be an open subset of W and $f : D \to Z$ a continuously differentiable function. Suppose that $f'(\mathbf{x}) : W \to Z$ is an isomorphism for all $\mathbf{x} \in D$. Show that $f(D)$ is an open subset of Z. In addition, if f is one-to-one on D, then show that f is a diffeomorphism on D.

Solution. If $\mathbf{y} \in f(D)$ then $by = f(\mathbf{x})$ for an $\mathbf{x} \in D$. The inverse function theorem, Theorem 6.1.4, shows that there is an open set $U_{\mathbf{x}}$ with $U_{\mathbf{x}} \subset D$ such that $f(U_{\mathbf{x}})$ is open subset of Z. Hence each $\mathbf{y} \in f(D)$ is contained in an open set $f(U_{\mathbf{x}}) \subset D$. Therefore $f(D)$ is an open set.

If f is one-to-one, then f has an inverse $g : f(D) \to D$. The inverse function is also a continuously differentiable function, again by the inverse function theorem. Hence f is a diffeomorphism between D and $f(D)$.

6.5 Let

$$\begin{aligned} A &= \{\, (x, y) \in \mathbb{R}^2 \mid x \in \mathbb{R}, 0 < y < \pi \,\}, \\ B &= \{\, (x, y) \in \mathbb{R}^2 \mid y > 0 \,\}. \end{aligned}$$

Define $f : A \to \mathbb{R}^2$ by

$$f(\mathbf{x}) = (e^x \cos y, e^x \sin y) \quad \text{for all } \mathbf{x} = (x, y) \in \mathbb{R}^2.$$

Show that f is a diffeomorphism from A onto B.

Solution. First, f is one-to-one on A. Suppose that $\mathbf{x} = (x, y), \mathbf{x}_1 = (x_1, y_1)$ are in A and $f(\mathbf{x}) = f(\mathbf{x}_1)$. Then $\|f(\mathbf{x})\| = \|f(\mathbf{x}_1)\|$ so that $e^x = e^{x_1}$. Hence, $x = x_1$. Thus, $\cos y = \cos y_1$ and $\sin y = \sin y_1$ so that $y = y_1$ (since y, y_1 are in $(0, \pi)$).

It is clear that $f(A) \subset B$. Let $(u, v) \in B$. Then there is an $r > 0$ and a θ, $0 < \theta < \pi$, such that $(u, v) = (r\cos\theta, r\sin\theta)$. Find $x \in \mathbb{R}$ such that $e^x = r$. Then $(x, \theta) \in A$ and $f(x, \theta) = (e^x\cos\theta, e^x\sin\theta) = (u, v)$. Hence, $f : A \to B$ is one-to-one. Finally, the Jacobian matrix of f is, for all $(x, y) \in A$,

$$\mathbf{J}f(x,\, y) = \begin{bmatrix} e^x\cos y & -e^x\sin y \\ e^x\sin y & e^x\cos y \end{bmatrix}.$$

Thus, $\det \mathbf{J}f(x,\, y) = e^{2x} \neq 0$ for all $\mathbf{x} \in A$. Hence $f'(x,\, y) : X \to Y$ is invertible for all $(x,\, y) \in A$. Then the inverse function theorem, Theorem 6.1.4, shows that f is a diffeomorphism from A onto B.

6.7 Define $f : \mathbb{R}^2 \to \mathbb{R}^2$ by $f(x,\, y) = (e^x + e^y, e^x - e^y)$. Show that f is a diffeomorphism. What is the range of f?

Solution. If $f(x,\, y) = (u,\, v)$, then $2e^x = u + v$ and $2e^y = u - v$. Hence $f : \mathbb{R}^2 \to \mathbb{R}^2$ is one-to-one. The Jacobian matrix of f is, for all $(x,\, y) \in \mathbb{R}^2$,

$$\mathbf{J}f(x,\, y) = \begin{bmatrix} e^x & e^y \\ e^x & -e^y \end{bmatrix}.$$

Thus, $\det \mathbf{J}f(x,\, y) = -2e^x e^y \neq 0$ for all $\mathbf{x} = (x, y) \in \mathbb{R}^2$. We see that f is a continuously differentiable function with an invertible derivative at every point in $\mathbb{R}^2$. Then Problem 6.3 shows that f is a diffeomorphism on $\mathbb{R}^2$. We see that the range of f consists of all $(u,\, v) \in \mathbb{R}^2$ such that $u + v > 0$ and $u - v > 0$. This an open quadrant in the uv-plane, bounded by the lines $u + v = 0$ and $u - v = 0$.

6.9 Repeat Problem 6.8 for

$$(u,\, v) = f(x,\, y) = (3x + 2y,\, 6x - 4y).$$

(Cf. Problem 1.31.)

Solution. We see that $f : \mathbb{R}^2 \to \mathbb{R}^2$ is an invertible linear transformation. Hence f is a diffeomorphism of $\mathbb{R}^2$ onto $\mathbb{R}^2$.

6.11 Repeat Problem 6.8 for

$$(u,\, v) = f(x,\, y) = ((x^2 + y^2)/(2x),\, (x^2 + y^2)/(2y)).$$

(Cf. Problem 1.33.)

Solution. We see that the domain of f is the set $A = \{\, (x,\, y) \mid xy \neq 0 \,\} \subset \mathbb{R}^2$. Also, as shown in the solution of Problem 1.33. We compute the derivative $f'(a,\, b) : \mathbb{R}^2 \to \mathbb{R}^2$ at $(x,\, y) \in A$. Let

$$u = (x^2 + y^2)/(2x) \;\; \text{and} \;\; v = (x^2 + y^2)/(2y).$$

Hence

$$\begin{aligned}\frac{\partial(u,\,v)}{\partial(x,\,y)} &= \frac{1}{2}\begin{bmatrix} 1-(y/x)^2 & 2(y/x) \\ 2(x/y) & 1-(x/y)^2 \end{bmatrix}. \text{ Therefore} \\ -2\det\frac{\partial(u,\,v)}{\partial(x,\,y)} &= 2+(y/x)^2+(x/y)^2 \neq 0 \text{ for all } (x,\,y)\in A.\end{aligned}$$

Hence $f : A \to \mathbb{R}^2$ is a diffeomorphism, since it is one-to-one on A and has an invertible derivative $f'(a,\,b) : \mathbb{R}^2 \to \mathbb{R}^2$ at every $(x,\,y) \in A$. Note that f maps the (open) first quadrant of the xy-plane (that is, the set $x > 0$ and $y > 0$) to the first quadrant of the uv-plane and similarly for the other quadrants.

6.13 Repeat Problem 6.8 for

$$(u,\,v) = f(x,\,y) = (p(x,\,y) + q(x,\,y),\, p(x,\,y) - q(x,\,y)),$$

where $p(x,\,y) = ((x+1)^2+y^2)^{1/2}$ and $q(x,\,y) = ((x-1)^2+y^2)^{1/2}$. (Cf. Problem 1.35.)

Solution. Here the domain of definition of f is the whole xy-plane. The solution of Problem 1.35 shows that f maps the open upper half-plane

$$A = \{\, (x,\,y) \mid y > 0 \,\}$$

of the xy-plane to the open strip

$$B = \{\, (u,\,v) \mid 2 < u \text{ and } -2 < v < 2 \,\}$$

in the uv-plane. We compute the derivative $f'(x,\,y) : \mathbb{R}^2 \to \mathbb{R}^2$ to see if f is a diffeomorphism on A. Let $u = p + q$ and $v = p - q$. We obtain

$$\begin{aligned}\det\frac{\partial(u,\,v)}{\partial(x,\,y)} &= \det\begin{bmatrix} (x+1)/p+(x-1)/q & y/p+y/q \\ (x+1)/p-(x-1)/q & y/p-y/q \end{bmatrix} \\ &= -4(y/pq) \neq 0 \text{ for all } (x,\,y)\in A.\end{aligned}$$

Hence f is a diffeomorphism between A and B. Note that f is also a diffeomorphism between the open lower half-plane $y < 0$ and B.

6.2 GRAPHS

6.15 Show that the linear tangent space to the graph of a function $f : \mathbb{R}^m \to \mathbb{R}^n$ at $(\mathbf{0}, f(\mathbf{0}))$ is the same as the linear tangent space to the graph of $y = f(\mathbf{x} - \mathbf{a})$ at the point where $\mathbf{x} = \mathbf{a}$.

Solution. Let $g(\mathbf{x}) = f(\mathbf{x} - \mathbf{a})$. The chain rule, Theorem 5.5.6, shows that $g'(\mathbf{a}) = f'(\mathbf{0})$. Hence the two linear tangent spaces described in the problem have the same equations $\mathbf{y} = f'(\mathbf{0})\mathbf{x}$ and $\mathbf{y} = g'(\mathbf{a})\mathbf{x}$.

6.17 Let Γ be the graph of $f : \mathbb{R}^m \to \mathbb{R}^n$. Let $\mathbf{a}, \mathbf{b}$ be in $\mathbb{R}^m$ such that $f'(\mathbf{b})^* f'(\mathbf{a}) = -I$. Show that the linear tangent space to Γ at $(\mathbf{a}, f(\mathbf{a}))$ and the linear tangent space to Γ at $(\mathbf{b}, f(\mathbf{b}))$ are orthogonal subspaces of $\mathbb{R}^{m+n}$.

Solution. Let M and N be linear tangent spaces to Γ at the point $(\mathbf{a}, f(\mathbf{a}))$ and $(\mathbf{b}, f(\mathbf{b}))$, respectively. Then

$$\begin{aligned} M &= \{ (\mathbf{x}, f'(\mathbf{a})\mathbf{x}) \mid \mathbf{x} \in \mathbf{R}^m \} \subset \mathbf{R}^{m+n} \\ N &= \{ (\mathbf{x}, f'(\mathbf{b})\mathbf{x}) \mid \mathbf{x} \in \mathbf{R}^m \} \subset \mathbf{R}^{m+n}. \end{aligned}$$

Let $\mathbf{w} = (\mathbf{x}, f'(\mathbf{a})\mathbf{x}) \in M$ and let $\mathbf{z} = (\mathbf{u}, f'(\mathbf{b})\mathbf{u}) \in N$. Then

$$\begin{aligned} \mathbf{w} \cdot \mathbf{z} &= \mathbf{x} \cdot \mathbf{u} + (f'(\mathbf{a})\mathbf{x}) \cdot (f'(\mathbf{b})\mathbf{u}) \\ &= \mathbf{x} \cdot \mathbf{u} + [f'(\mathbf{b})^* f'(\mathbf{a})\mathbf{x}] \cdot \mathbf{u} \\ &= \mathbf{x} \cdot \mathbf{u} + (-I\mathbf{x}) \cdot \mathbf{u} = \mathbf{x} \cdot \mathbf{u} - \mathbf{x} \cdot \mathbf{u} = 0. \end{aligned}$$

6.19 The linear tangent space to the graph of a function $f : \mathbb{R}^n \to \mathbb{R}$ at any point $(\mathbf{a}, f(\mathbf{a}))$ is a subspace of $\mathbb{R}^{n+1}$. Show that this subspace has dimension n.

Solution. The linear tangent space to the graph of $f : \mathbf{R}^n \to \mathbf{R}^m$ is the subspace $M = \{ (\mathbf{x}, f'(\mathbf{a})\mathbf{x}) \mid \mathbf{x} \in \mathbf{R}^n \}$ of $\mathbf{R}^{m+n}$. Let $\mathbf{e}_1, \ldots, \mathbf{e}_n$ be the standard basis vectors of $\mathbf{R}^n$. Then M contains the the set $S = \{(\mathbf{e}_1, f'(\mathbf{a})\mathbf{e}_1), \ldots, (\mathbf{e}_n, f'(\mathbf{a})\mathbf{e}_n)\}$, which is clearly linearly independent. Hence, $\dim M \geq n$. We now show that S spans M. Every vector $\mathbf{m}$ in M is of the form

$$\mathbf{m} = (\mathbf{u}, f'(\mathbf{a})\mathbf{u}) \quad \text{for some } \mathbf{u} = (u_1, \ldots, u_n) \in \mathbf{R}^n.$$

Hence,

$$\begin{aligned} \mathbf{m} &= (u_1\mathbf{e}_1 + \cdots + u_n\mathbf{e}_n, f'(\mathbf{a})(u_1\mathbf{e}_1 + \cdots + u_n\mathbf{e}_n)) \\ &= (u_1\mathbf{e}_1 + \cdots + u_n\mathbf{e}_n, u_1 f'(\mathbf{a})\mathbf{e}_1 + \cdots + u_n f'(\mathbf{a})\mathbf{e}_n) \\ &= (u_1\mathbf{e}_1, u_1 f'(\mathbf{a})\mathbf{e}_1) + \cdots + (u_n\mathbf{e}_n, u_n f'(\mathbf{a})\mathbf{e}_n) \\ &= u_1(\mathbf{e}_1, f'(\mathbf{a})\mathbf{e}_1) + \cdots + u_n(\mathbf{e}_n, f'(\mathbf{a})\mathbf{e}_n). \end{aligned}$$

Thus, $\mathbf{m}$ is in the span of S, as to be shown.

6.3 MANIFOLDS IN PARAMETRIC REPRESENTATIONS

6.21 Let $M = \{ (\mathbf{x}, (x^2 + y - z, y^2 + x + z)) \mid \mathbf{x} = (x, y, z) \in \mathbb{R}^3 \}$. Is M a manifold in $\mathbb{R}^5$? If it is, find a parametric equation for M and an underlying diffeomorphism.

Solution. The set M is a manifold because it is the graph of the function

$$f(x,\, y,\, z)) = (x^2 + y - z,\, y^2 + x + z), \quad (x,\, y,\, z) \in \mathbb{R}^3.$$

This is a function with the domain $\mathbb{R}^3$ and the range space $\mathbb{R}^2$. Its graph is a subset of $\mathbb{R}^3 \times \mathbb{R}^2 \simeq \mathbb{R}^5$. A parametric equation for M is

$$\varphi(x,\, y,\, z) = (x,\, y,\, z,\, x^2 + y - z,\, y^2 + x + z),$$

which is a function $\varphi : \mathbb{R}^3 \to \mathbb{R}^5$. The underlying diffeomorphism $\Phi : \mathbb{R}^5 \to \mathbb{R}^5$ is

$$\Phi(x,\, y,\, z,\, u,\, v) = (x,\, y,\, z,\, u + x^2 + y - z,\, v + y^2 + x + z).$$

The parametric equation for M is obtained by restricting $\Phi : \mathbb{R}^5 \to \mathbb{R}^3$ to $\mathbb{R}^3$.

6.23 Let $M \subset \mathbb{R}^n$ be an r-manifold and let $N \subset \mathbb{R}^m$ be an s-manifold. Show that $M \times N \subset \mathbb{R}^m \times \mathbb{R}^n = \mathbb{R}^{m+n}$ is an $(r+s)$-manifold.

Solution. Let $(\mathbf{a},\, \mathbf{b}) \in M \times N \subset \mathbb{R}^m \times \mathbb{R}^n$, with $M \subset \mathbb{R}^m$ and $N \subset \mathbb{R}^n$. Definition 6.3.1 of manifolds shows that there is an open set $G \subset \mathbb{R}^m$ containing $\mathbf{a}$ such that $G \cap M$ is a graph in $\mathbb{R}^m$, with respect to a coordinate system $(X,\, Y)$ in $\mathbb{R}^m$. Similarly there is an open set $H \subset \mathbb{R}^n$ containing $\mathbf{b}$ such that $H \cap N$ is a graph in $\mathbb{R}^n$, with respect to a coordinate system $(U,\, V)$ in $\mathbb{R}^n$. Then we see that $(X \times U,\, Y \times V)$ is a coordinate system in $\mathbb{R}^m \times \mathbb{R}^n$ and $(G \times H) \cap (M \times N)$ is a graph in $\mathbb{R}^m \times \mathbb{R}^n$ with respect to this coordinate system. But $G \times H$ is an open set in $\mathbb{R}^m \times \mathbb{R}^n$ containing $(\mathbf{a},\, \mathbf{b})$. Hence $M \times N$ is a manifold in $\mathbb{R}^m \times \mathbb{R}^n$. Also, $\dim X = r$ and $\dim U = s$. Therefore $\dim(X \times U) = r + s$. Hence $M \times N$ is an $(r+s)$-dimensional manifold.

6.25 Let C be an open subset of U and $\varphi : C \to Z$ a one-to-one $\mathcal{C}^1$ function with a one-to-one derivative at every point. Example 6.3.13 shows that $\varphi(C)$ does not have to be a manifold. Let C_0 be an open subset of C with closure $\overline{C_0} \subset C$. Show that $\varphi(C_0)$ is a manifold.

Solution. Note that the restriction of $\varphi : C \to Z$ to any set $A \subset C$ is a continuous and invertible function. The inverse function is not necessarily continuous. This was shown in Example 4.4.18. Other examples can be obtained from Example 6.3.13. However, apply Theorem 4.4.21 to conclude that the restriction of φ to the compact set $\overline{C_0}$ has a continuous inverse. Therefore the restriction of this inverse function to $\varphi(C_0)$ is a continuous function $\psi : \varphi(C_0) = M \to C_0$. Let $\mathbf{m} \in M$ and let $\mathbf{c} = \psi(\mathbf{m})$. Theorem 6.3.11 shows that $\mathbf{c}$ has a neighborhood $C_1 \subset C_0$ such that $\varphi(C_1)$ is a manifold. Use the continuity of $\psi : \varphi(C_0) \to C_0$ to conclude that there is an open set $G \subset Z$ such that $\mathbf{m} \in G$ and such that $\psi(G \cap M) \subset C_1$. This implies that $G \cap M \subset \varphi(C_1)$. Hence $G \cap M$ is a manifold. Therefore M is a manifold. Example 6.3.13 shows that this is false without the continuity of the inverse function $\psi : M \to C_0$.

6.27 Let C be an open subset of U and $\varphi : C \to Z$ a one-to-one $\mathcal{C}^1$ function with a one-to-one derivative at every point. Assume that the inverse function $\psi : \varphi(C) \to C$ is continuous. Show that $\varphi(C)$ is a manifold in Z.

Solution. This is the same as the solution of Problem 6.25. Let $\mathbf{m} \in M = \varphi(C)$. If $\mathbf{m} \in M$ then $\psi(\mathbf{m}) = \mathbf{c} \in C$ has a neighborhood C_1 such that $\varphi(C_1)$ is a manifold. We know that $\mathbf{m} \in \varphi(C_1)$, but to show that $M = \varphi(C)$ is a manifold we need an open set $G \subset Z$ such that $\mathbf{m} \in G$ and such that $G \cap M \subset \varphi(C_1)$. This is obtained by the continuity of $\psi : M \to C$, as in the solution of Problem 6.25.

6.4 MANIFOLDS IN IMPLICIT REPRESENTATIONS

6.29 Let $F : \mathbb{R}^3 \to \mathbb{R}^2$ be given by $F(x, y, z) = (y^2 + z, z + x)$ for all $\mathbf{x} = (x, y, z) \in \mathbb{R}^3$. Show that

$$M = \{ (x, y, z) \in \mathbb{R}^3 \mid F(\mathbf{x}) = (0, 0) \}$$

is a manifold in $\mathbb{R}^3$. Find the tangent space of M at $\mathbf{m} = (1, 1, -1)$. Also, find the normal space of M at the same point $\mathbf{m}$.

Solution. We see that

$$F'(a, b, c)(x, y, z) = (2by + z, x + z).$$

Hence $F'(a, b, c)$ maps $(1, 0, 0)$ to $(0, 1)$ and $(0, 0, 1)$ to $(1, 0)$. This shows that $F'(a, b, c) : \mathbb{R}^3 \to \mathbb{R}^2$ maps $\mathbb{R}^3$ onto $\mathbb{R}^2$, for all $(a, b, c) \in \mathbb{R}^3$. Then Theorem 6.4.1 shows that $F(x, y, z) = (y^2 + z, z + x) = (0, 0)$ is the implicit equation of a manifold M. In this case this is a curve. We see that $\mathbf{m} = (1, 1, -1) \in M$. The tangent space of M at $\mathbf{m}$ is the kernel of

$$F'(1, 1, 1)(x, y, z) = (2y + z, x + z).$$

This kernel is a one-dimensional space. The vector $(2, 1, -2)$ is a nonzero vector in this space. Hence a parametric equation of the affine tangent line is

$$(x, y, z) = (1, 1, -1) + t(2, 1, -2), \quad t \in \mathbb{R}.$$

The normal space is the orthogonal complement of the tangent space. Hence an equation of the liner normal plane is

$$2x + y - 2z = 0.$$

An equation of the affine normal space is

$$2(x - 1) + (y - 1) - 2(z + 1) = 0 \text{ or } 2x + y - 2z = 5.$$

6.31 Consider the system

$$\begin{aligned} x^2 + \frac{1}{2}y^2 + z^3 - z^2 &= \frac{3}{2} \\ x^3 + y^3 - 3y + z &= -3. \end{aligned}$$

Can we solve for y and z as a function of x for (x, y, z) in a neighborhood of $(-1, 1, 0)$?

Solution. Let $F_1(\mathbf{x}) = x^2 + \frac{1}{2}y^2 + z^3 - z^2 - \frac{3}{2}$, $F_2(\mathbf{x}) = x^3 + y^3 - 3y + z + 3$ and let $F(\mathbf{x}) = (F_1(\mathbf{x}), F_2(\mathbf{x}))$ for all $\mathbf{x} = (x, y, z) \in \mathbf{R}^3$. Then $F : \mathbf{R}^3 \to \mathbf{R}^2$ have continuously differentiable components and $F(-1, 1, 0) = (0, 0)$. Also,

$$\det \begin{bmatrix} (\partial F_1/\partial y)(-1,1,0) & (\partial F_1/\partial z)(-1,1,0) \\ (\partial F_2/\partial y)(-1,1,0) & (\partial F_2/\partial z)(-1,1,0) \end{bmatrix} = \det \begin{bmatrix} 1 & 0 \\ -3 & 1 \end{bmatrix} \neq 0.$$

Hence, by the implicit function theorem, Theorem 6.4.6, as in Example 6.4.7, there is a neighborhood A of -1 in $\mathbf{R}$ and a continuously differentiable function $f : A \to \mathbf{R}^2$ such that $f(-1) = (1, 0)$ and $F(x, f(x)) = (0, 0)$ for all $x \in A$. Thus, if f_1, f_2 are the component functions of f, then $F(x, (f_1(x), f_2(x)) = (0, 0)$ for all $x \in A$. Thus, $y = f_1(x), z = f_2(x)$ for each solution (x, y, z) to the given system that lies in a neighborhood of $(-1, 1, 0)$.

6.33 Show that the system

$$\begin{aligned} xv + yu &= 1 \\ xy &= uv \end{aligned}$$

defines $(u,\, v) = h(x,\, y)$ implicitly as a function of $(x,\, y)$ for $(x,\, y)$ in a neighborhood of $(1, 0)$. Compute $h'(1, 0)$.

Solution. Here, $F_1(x, y, u, v) = xv + yu - 5$, $F_2(x, y, u, v) = xy - uv - 5$ for all (x, y) and (u, v) in $\mathbf{R}^2$. Also, $\mathbf{a} = (1, 2)$ and $\mathbf{b} = (3, -1)$. Hence,

$$\begin{bmatrix} (\partial F_1/\partial u)(\mathbf{a},\mathbf{b}) & (\partial F_1/\partial v)(\mathbf{a},\mathbf{b}) \\ (\partial F_2/\partial u)(\mathbf{a},\mathbf{b}) & (\partial F_2/\partial v)(\mathbf{a},\mathbf{b}) \end{bmatrix}^{-1} = \begin{bmatrix} 2 & 1 \\ 1 & 3 \end{bmatrix}^{-1} = \frac{1}{5}\begin{bmatrix} 3 & -1 \\ -1 & 2 \end{bmatrix}.$$

Also,

$$\begin{bmatrix} (\partial F_1/\partial x)(\mathbf{a},\mathbf{b}) & (\partial F_1/\partial y)(\mathbf{a},\mathbf{b}) \\ (\partial F_2/\partial x)(\mathbf{a},\mathbf{b}) & (\partial F_2/\partial y)(\mathbf{a},\mathbf{b}) \end{bmatrix} = \begin{bmatrix} -1 & 3 \\ 2 & 1 \end{bmatrix}.$$

Hence,

$$f'(1,2) = f'(\mathbf{a}) = -\frac{1}{5}\begin{bmatrix} 3 & -1 \\ -1 & 2 \end{bmatrix}\begin{bmatrix} -1 & 3 \\ 2 & 1 \end{bmatrix} = \begin{bmatrix} 1 & -\frac{8}{5} \\ -1 & \frac{1}{5} \end{bmatrix}.$$

6.5 DIFFERENTIATION ON MANIFOLDS

6.35 Find the maximum and minimum values of $x^2+y^2+z^2$ given that $x+y+z=0$ and $(x-3)^2+y^2+z^2=9$.

Solution. Let $f(\mathbf{x})=x^2+y^2+z^2$, $g(\mathbf{x})=(x-3)^2+y^2+z^2-9$ and $h(\mathbf{x})=x+y+z$. Then we want to find the maximum value of f on the sets $g=h=0$. Thus, we seek real λ, μ, and $\mathbf{x}$ such that $\nabla f(\mathbf{x})=\lambda\nabla g(\mathbf{x})+\mu\nabla h(\mathbf{x})$, and $g(\mathbf{x})=h(\mathbf{x})=0$. Thus,

$$\begin{aligned} 2x &= 2\lambda(x-2)+\mu \\ 2y &= 2y\lambda+\mu \\ 2z &= 2z\lambda+\mu \\ (x-3)^2+y^2+z^2 &= 9 \\ x+y+z &= 0. \end{aligned}$$

The second and third equations imply that

$$(y-z)(1-\lambda)=0.$$

Hence, $y=z$ or $\lambda=1$. If $\lambda=1$, then the second equation gives $\mu=0$. But if $\mu=0$, then the first equation leads to a contradiction. Hence, $\lambda\neq 1$. Therefore, $y=z$. From the last equation, we get $x=-2z$. Hence, the fourth equation now gives $(-2z-3)^2+z^2+z^2=9$ so that $6z^2+12z=0$. Thus, $z=0$ or $z=-2$. So,

$$z=0,\ y=0,\ x=0,\ \mu=0,\ \lambda=0$$

or

$$z=-2,\ y=-2,\ x=4,\ \mu=2,\ \lambda=\frac{3}{2}.$$

It is now clear that the maximum value of f on $g=h=0$ occurs at $(4,-2,-2)$ and the minimum value occurs at $(0,0,0)$.

6.37 Find the maximum value of xyz where $x^2+y^2+z^2=3$.

Solution. Let $f(x,y,z)=xyz$ and let $S=\{\mathbf{x}=(x,y,z)\in\mathbf{R}^3 \mid g(\mathbf{x})=3\}$, where $g(x,y,z)=x^2+y^2+z^2$. We seek $\lambda\in\mathbf{R}$ and $\mathbf{x}=(x,y,z)\in S$ such that $\nabla f(\mathbf{x})=\lambda\nabla g(\mathbf{x})$. Now, $\nabla f(\mathbf{x})=\lambda\nabla g(\mathbf{x})$ if and only if

$$\begin{aligned} yz &= 2\lambda x \\ xz &= 2\lambda y \\ xy &= 2\lambda z. \end{aligned}$$

Multiply the first equation by x, the second equation by y and the third equation by z give

$$\lambda x^2=\lambda y^2=\lambda z^2.$$

Since the maximum value on S cannot be 0, $\lambda \neq 0$ (or else the very first equation above would imply that $yz = 0$ so that $f(x, y, z) = 0$). Hence, $x^2 = y^2 = z^2$. Thus, since $x^2 + y^2 + z^2 = 3$, we have $3x^2 = 3$ so that $x^2 = 1$. Hence, $y^2 = z^2 = 1$. That is, $x = \pm 1, y = \pm 1, z = \pm 1$.It is clear that the maximum value of f on S occurs when $x = 1 = y = z$ and this maximum value is 1.

6.39 Find the minimum value of $5x - 2y + 7z$ where $x^2 + 2y + 4z^2 = 9$

Solution. Let $f(\mathbf{x}) = 5x - 2y + 7z$ and let $S = \left\{ \mathbf{x} \in \mathbf{R}^3 \mid x^2 + 2y^2 + 4z^2 = 9 \right\}$. Suppose that the minimum value of f on S occurs at $\mathbf{x} \in S$. Then there is some $\lambda \in \mathbf{R}$ such that $\nabla f(\mathbf{x}) = \lambda \nabla g(\mathbf{x})$. That is,

$$\begin{aligned} 5 &= 2\lambda x \\ -2 &= 2\lambda y \\ 7 &= 8\lambda z. \end{aligned}$$

Necessarily, $\lambda \neq 0$. Since $x^2 + 2y^2 + x^2 = 9$, we then have

$$\left(\frac{5}{2\lambda}\right)^2 + 2\left(-\frac{1}{\lambda}\right)^2 + 4\left(\frac{7}{8\lambda}\right)^2 = 9.$$

Hence, $\lambda^2 = 181/144$. Hence,

$$\lambda = \pm \frac{\sqrt{181}}{12}.$$

It should be clear that the minimum value of f on S occurs for

$$\lambda = -\frac{\sqrt{181}}{12},$$

at the point where

$$x = \frac{5}{2\lambda}, \quad y = -\frac{1}{\lambda}, \quad z = \frac{7}{8\lambda}.$$

6.41 Find the points on the curve $x^2 + xy + y^2 = 3$ closest to and farthest from the origin.

Solution. Let $f(\mathbf{x}) = x^2 + y^2$ for all $\mathbf{x} = (x, y) \in \mathbf{R}^3$. Let $S = \left\{ \mathbf{x} \in \mathbf{R}^2 \mid g(\mathbf{x}) = 3 \right\}$, where $g(\mathbf{x}) = x^2 + xy + y^2$. We want to find points in S that will maximize f and points in S that will minimize f. Thus, we seek $\lambda \in \mathbf{R}$ and $\mathbf{x} \in S$ such that $\nabla f(\mathbf{x}) = \lambda \nabla g(\mathbf{x})$. Thus,

$$\begin{aligned} 2x &= \lambda(2x + y) \\ 2y &= \lambda(x + 2y). \end{aligned}$$

Multiply the first equation by y, multiply the second equation by x, then subtract, we have

$$\lambda(y^2 - x^2) = 0.$$

If $\lambda = 0$, then $x = 0 = y$, an impossibility , since $\mathbf{x} \in S$. Hence, $\lambda \not= 0$. Thus, $x^2 = y^2$ so that $x = \pm y$. If $x = y$, then $x^2 + xy + x^2 = 3 \iff x^2 = 1$ so that $x = \pm 1$. If $x = -y$, then $x^2 + xy + y^2 = 3 \iff x^2 = 3$ so that $x = \pm\sqrt{3}$. If $x = \pm 1$, then $f(\mathbf{x}) = x^2 + y^2 = 2$. If $x = \pm\sqrt{3}$, then $f(\mathbf{x}) = x^2 + y^2 = 6$. Thus, the points $(1, 1), (-1, -1)$ on the curve $x^2 + xy + y^2 = 3$ are closest from the origin; the points $(\sqrt{3}, -\sqrt{3}), (-\sqrt{3}, \sqrt{3})$ on the curve $x^2 + xy + y^2 = 3$ are farthest from the origin.

CHAPTER 7

HIGHER-ORDER DERIVATIVES

7.1 DEFINITIONS

7.2 CHANGE OF ORDER IN DIFFERENTIATION

7.3 SEQUENCES OF POLYNOMIALS

7.4 LOCAL EXTREMAL VALUES

7.1 Show that the polynomial $P(x, y) = Ax^2 + 2Bxy + Cy^2$ is positive definite if and only if $B^2 < AC$ and $0 < A$.

Solution. We see that $P : \mathbb{R}^2 \to \mathbb{R}$ is a real valued homogeneous polynomial of degree two. It is positive definite if $P(x, y) > 0$ whenever $(x, y) \neq (0, 0)$. This is

Analysis in Vector Spaces.
By M. A. Akcoglu, P. F. A. Bartha and D. M. Ha

given in Definition 7.4.1. Assume that P is positive definite. Then

$$P(x,\,1) = Ax^2 + 2Bx + C > 0 \ \text{ for all } \ x \in \mathbb{R}.$$

Therefore the the equation $Ax^2 + 2Bx + C = 0$ has no real roots. This means that $B^2 < AC$. Also $P(0,\,1) = C > 0$. Then $A > 0$ since $0 \le B^2 < AC$.

Conversely assume that $B^2 < AC$ and $0 < A$. Then also $0 < C$. Hence

$$At^2 + 2Bt + C > 0 \ \text{ for all } \ t \in \mathbb{R} \ \text{ and} \tag{7.1}$$

$$A + 2Bs + Cs^2 > 0 \ \text{ for all } \ t \in \mathbb{R}. \tag{7.2}$$

Assume that $P(x,\,y) \le 0$ for some $(x,\,y) \ne (0,\,0)$. If $y \ne 0$ then (7.1) is violated with $t = (x/y)$. If $x \ne 0$ then (7.2) is violated with $s = (y/x)$. Hence we must have $P(x,\,y) > 0$ for all $(x,\,y) \ne (0,\,0)$.

PART III

INTEGRATION

CHAPTER 8

MULTIPLE INTEGRALS

8.1 JORDAN SETS AND VOLUME

8.1 Find $I_k(E)$ and $O_k(E)$ for $k = 0, 1, 2, 3$, where

$$\begin{aligned}
&(a) \qquad E = \{\, (x, y) \mid 0 \leq x,\ 0 \leq y,\ x^2 + y^2 \leq 1 \,\} \subset \mathbb{R}^2, \\
&(b) \qquad E = \{\, (x, y) \mid 0 < x,\ 0 < y,\ x^2 + y^2 < 1 \,\} \subset \mathbb{R}^2, \\
&(c) \qquad E = \{\, (x, y) \mid 0 \leq x,\ 0 \leq y,\ x + y \leq 1 \,\} \subset \mathbb{R}^2, \\
&(d) \qquad E = \{\, (x, y) \mid 0 < x,\ 0 < y,\ x + y < 1 \,\} \subset \mathbb{R}^2.
\end{aligned}$$

Solution. Recall that $I_k(E)$ and $O_k(E)$ are the inner and outer approximations of E. That is, $I_k(E)$ is the union of all k th order squares contained in E, while $O_k(E)$ is the union of all such squares that meet E. Notice that, if E is any of the four sets above, a k th order square will lie in $I_k(E)$ iff all four vertices lie in E, and in $O_k(E)$ iff at least one vertex lies in E. So the problem is mainly to determine which of the relevant vertices lie in E. For $k = 0, 1, 2, 3$, the relevant possibilities are vertices

Analysis in Vector Spaces.
By M. A. Akcoglu, P. F. A. Bartha and D. M. Ha

(x, y) where x and y range from 0 to 2 in increments of 2^{-k}. We obtain the solutions below by calculation for $k = 0$ and $k = 3$; the cases $k = 1$ and $k = 2$ are left to the reader.

(a) Here, E is a quadrant of the unit circle. For $k = 0$, $I_0(E) = \emptyset$, while

$$O_0(E) = [0, 1) \times [0, 1) \cup [1, 2) \times [0, 1) \cup [0, 1) \times [1, 2).$$

Note that each of the latter two sets intersects E at just one point. For $k = 3$, we describe the approximations in terms of successive columns of $3rd$ order squares (side length $1/8$) stacked above the intervals $[0, 1/8), \cdots, [1, 9/8)$. $I_3(E)$ consists of 7 squares above $[0, 1/8)$, 7 above $[1/8, 1/4)$, 7 above $[1/4, 3/8)$, 6 above $[3/8, 1/2)$, 6 above $[1/2, 5/8)$, 5 above $[5/8, 3/4)$, and 3 above $[3/4, 7/8)$. Let's represent this as the sequence $< 7, 7, 7, 6, 6, 5, 3 >$ for a total of 41 squares of side $1/8$. Employing the same notation, $O_3(E)$ may be represented, as $< 9, 8, 8, 8, 7, 7, 6, 4, 1 >$ for a total of 58 squares.

(b) For $k = 0$, $I_0(E) = \emptyset$; $O_0(E) = [0, 1) \times [0, 1)$. For $k = 3$, using the notation above, we represent $I_3(E)$ just as in part (a), while $O_3(E)$ may be represented as $< 8, 8, 8, 8, 7, 7, 6, 4 >$ for a total of 56 squares.

(c) Here, E is the triangular region bounded by the axes and the line $x + y \leq 1$. Since kth order squares are open along the top and right edges, we have the following results. For $k = 0$, $I_0(E)$ and $O_0(E)$ are just as in part (a). For $k = 3$, using the notation above, we represent $I_3(E)$ by the sequence $< 7, 6, 5, 4, 3, 2, 1 >$ for a total of 28 squares, and $O_3(E)$ by $< 9, 8, 7, 6, 5, 4, 3, 2, 1 >$ for a total of 45 squares.

(d) For $k = 0$, $I_0(E)$ and $O_0(E)$ are just as in part (b). For $k = 3$, we represent $I_3(E)$ just as in part (c), while $O_3(E)$ may be represented as $< 8, 7, 6, 5, 4, 3, 2, 1 >$ for a total of 36 squares.

8.3 For each $n \in \mathbb{N}$ find a negligible set E_n such that $E = \cup_{n \in \mathbb{N}}$ is not negligible.

Solution. The set of rational numbers in $[0, 1]$ is not negligible, but it is enumerable as a list: $r_1, r_2, \cdots$. Put $E_n = \{r_n\}$.

8.5 Let $E = \{ (x, y) \mid |x| + |y| \leq 1 \} \subset \mathbb{R} \times \mathbb{R}$. Find $V_E(J)$ for any interval J.

Solution. E is the diamond-shaped region whose four vertices are $(1, 0)$, $(0, 1)$, $(-1, 0)$ and $(0, -1)$. Let J be any interval $[a, b]$ along the y-axis, where $a < b$. Clearly, if $1 < a$ or $b < -1$ then each cross-section E_y is empty, and $V_E(J) = 0$. If $-1 \leq b \leq 0$, then $V_E(J) = 2 - 2|b|$ if $a \leq -1$, since in this case the union of cross-sections over J is the interval $[-(1 - |b|), (1 - |b|)]$ and the intersection is empty; by contrast, if $-1 < a \leq b$ then $V_E(J) = 2(|a| - |b|)$ because the intersection of cross-sections over J is now $[-(1 - |a|), (1 - |a|)]$.

By similar reasoning, if $0 \leq a \leq 1$, then $V_E(J) = 2 - 2|a|$ if $b \geq 1$ and $V_E(J) = 2(|b| - |a|)$ if $a \leq b < 1$.

If either $a < -1$ and $b \geq 0$, or $a \leq 0$ and $b > 1$, $V_E(J) = 2$. The only remaining case is where $-1 \leq a \leq 0 \leq b \leq 1$. In this case, reasoning similar to the above cases shows that $V_E(J) = 2 \cdot max(|a|, |b|)$.

8.7 Let E be the bounded region in the xy-plane between the parabolas $y = x^2$ and $y = 2x^2 - 1$. Find the variations of the cross-sections of E. Consider the cross-sections both with the vertical $x =$ constant lines and with the horizontal $y =$ constant lines.

Solution. We provide solutions for cross-sections with horizontal $y =$ constant lines. Let J be an interval $[a, b]$ along the y-axis, where $a < b$. If $-1 \leq k \leq 0$, then the cross-section through $y = k$ is the interval $[-\sqrt{(k+1)/2}, \sqrt{(k+1)/2}]$. If $0 < k < 1$, then a cross-section through $y = k$ consists of the intervals $[-\sqrt{(k+1)/2}, -\sqrt{k}]$ and $[\sqrt{k}, \sqrt{(k+1)/2}]$.

Suppose $a \leq -1$. If $b < -1$, then $V_E(J) = 0$. If $-1 \leq b \leq 1$, then the union of cross-sections is the interval $[-\sqrt{(b+1)/2}, \sqrt{(b+1)/2}]$ and the intersection is empty, so $V_E(J) = 2 \cdot \sqrt{(b+1)/2}$. If $b > 1$, then $V_E(J) = 2$.

Suppose next that $-1 < a < 0$. If $a < b < 0$, then $V_E(J) = 2 \cdot (\sqrt{(b+1)/2} - \sqrt{(a+1)/2})$. If $0 \leq b \leq (a+1)/2$, then the union of the cross-sections is $[-\sqrt{(b+1)/2}, \sqrt{(b+1)/2}]$, but the intersection is now $[-\sqrt{(a+1)/2}, -\sqrt{b}]$ plus $[\sqrt{b}, \sqrt{(a+1)/2}]$. So we have $V_E(J) = 2 \cdot [\sqrt{(b+1)/2} - (\sqrt{(a+1)/2} - \sqrt{b})]$. If $(a+1)/2 < b \leq 1$, then the intersection of cross-sections is empty and $V_E(J) = 2 \cdot \sqrt{(b+1)/2}$. Finally, if $b > 1$, then $V_E(J) = 2$.

Suppose next that $0 \leq a \leq 1$. By reasoning similar to the above, if $a < b \leq (a+1)/2$ we get $V_E(J) = 2 \cdot [\sqrt{(b+1)/2} - \sqrt{a}] - 2 \cdot [\sqrt{(a+1)/2} - \sqrt{b}]$. If $(a+1)/2 < b \leq 1$, then $V_E(J) = 2 \cdot [\sqrt{(b+1)/2} - \sqrt{a}]$. If $b > 1$, then $V_E(J) = 2 \cdot [1 - \sqrt{a}]$.

Finally, if $a > 1$, then obviously $V_E(J) = 0$.

8.9 Let E be the bounded region in the xyz-space $\mathbb{R}^2 \times \mathbb{R}$ between the cylinders $x^2 + z^2 = 1$ and $y^2 + z^2 = 1$. Find the variation of the cross-sections of E with the horizontal $z =$ constant planes.

Solution. First, each cross-section with $z = k$ is the square of side $2 \cdot \sqrt{1 - k^2}$. The area of this square is thus $4 \cdot (1 - k^2)$. For any interval $J = [a, b]$ along the z-axis, we have to consider cases.

If $a > 1$ or $b < -1$, so that J does not even meet E, then $V_E(J) = 0$.

Suppose $0 \leq a \leq 1$. If $b > 1$, then the union of cross-sections is just the cross-section at $z = a$ and the intersection is empty, so $V_E(J) = 4 \cdot (1 - a^2)$. If $a \leq b \leq 1$, then the union is the same but the intersection is the cross-section at $z = b$, so that $V_E(J) = 4 \cdot [(1 - a^2) - (1 - b^2)]$.

Suppose $-1 \leq a \leq 0$. If $b > 1$, then the union of cross-sections is the largest cross-section at $z = 0$ but the intersection is empty, so $V_E(J) = 4$. If $|a| \leq b \leq 1$, then $V_E(J) = 4 - 4 \cdot (1 - b^2)$. If $0 \leq b \leq |a|$, then $V_E(J) = 4 - 4 \cdot (1 - a^2)$. Finally, if $a \leq b < 0$, then the union of cross-sections is the square at $z = b$ and the intersection is the square at $z = a$, so that $V_E(J) = 4 \cdot [(1 - b^2) - (1 - a^2)]$.

The remaining case, where $a < -1$, is omitted.

8.11 Show that both $|x| + |y| \leq 1$ and $x^2 + y^2 \leq 1$ are Jordan sets in $\mathbb{R}^2$.

Solution. We provide the solution for the set E defined by $|x| + |y| \leq 1$. From problem 8.5, it is clear that if $J = [a, b]$ is an interval on the y-axis, then $V_E(J)$ depends continuously on a and b. So E has a continuously changing cross-section on J. By problem 8.10, it follows that E is a Jordan set. A similar argument for $x^2 + y^2 \leq 1$ relies upon problem 8.6.

8.13 Cylinders. Let A be a Jordan set in $\mathbb{R}^n$ and $\mathbf{c} = (\mathbf{a}, h) \in \mathbb{R}^n \times \mathbb{R}$ with $h \neq 0$. Then the set

$$C = C(A, \mathbf{c}) = \{ (\mathbf{x}, 0) + t(\mathbf{a}, h) \mid \mathbf{x} \in A, \ 0 \leq t \leq 1 \} \subset \mathbb{R}^{n+1}$$

is called a cylinder. Show that all cylinders are Jordan sets.

Solution. First suppose A is the unit square in R^2. Then each cross-section of the cylinder C is a unit square of the form $A + t\mathbf{a}$, with $0 \leq t \leq 1$. If we fix $0 < t < 1$, then variation of cross-sections of C over $(t - \delta, t + \delta)$ can be made as small as we please. For the union of these cross-sections lies in a square whose edge has length $(1 + 2\delta||\mathbf{a}||)$, while the intersection contains a square whose edge has length $(1 - 2\delta||\mathbf{a}||)$. The difference between the areas of the outer approximating square and the inner approximating square can be made small by taking δ suitably small. For this special case, then, problem 8.10 shows that C is a Jordan set.

This argument can be extended to the case where A is any cube in $\mathcal{C}^n_k$. We can then extend it to any finite union of cubes because any cross-section of the cylinder based on a union of cubes is just the union of the cross-sections of cylinders based on each cube. Finally, if A is a Jordan set in R^n, then we can find inner and outer approximations F and G for A by unions of cubes in $\mathcal{C}^n_k$. The cross-sections of the cylinder based on A are then approximated by cross-sections for the cylinders based on F and G, from which it follows that variations of these cross-sections is continuous and E is a Jordan set.

8.15 Show that all triangles in $\mathbb{R}^2$ are Jordan sets.

Solution. If E is a triangle with one side along the x-axis, then E is a cone in the sense of problem 8.14 and therefore a Jordan set. Otherwise, every horizontal cross-section of E will be an interval along the x-axis determined by two lines of the form $y = bx$ and $y = cx$. If we fix a small interval J along the y-axis, then it is clear that the endpoints of the horizontal cross-sections E_y for $y \in J$ vary continuously, and hence that the variation over these cross-sections is continuous. Hence E is a Jordan set.

8.17 Show that all Euclidean balls

$$E = \left\{ \mathbf{x} = (x_1, \ldots, x_n) \in \mathbb{R}^n \,\middle|\, \sum_i (x_i - c_i)^2 \le R^2 \right\}$$

are Jordan sets in $\mathbb{R}^n$. Here $\mathbf{c} = (c_1, \ldots, c_n)$ is the center of the ball. (Hint. First assume that $\mathbf{c} = \mathbf{0}$ and proceed by induction on $n \in \mathbb{N}$.)

Solution. We can assume $\mathbf{c} = \mathbf{0}$ without loss of generality, because any translation of a Jordan set is a Jordan set. (To see this, transfer the inner and outer approximations by cubes.) Also assume that $R = 1$.

If $n = 1$ then E is the interval $[-1, 1]$, and hence a Jordan set. Assume the result for a certain $n \in \mathbb{N}$. Let a_n be the n dimensional volume of the unit ball $B_1^n(\mathbf{0}) \subset \mathbb{R}^n$. Then the volume of the ball of radius r is $v^n(B_r^n(\mathbf{0})) = r^n a_n$ for $r > 0$. To see this note that the scaling transformation $\mathbb{R}^n \to \mathbb{R}^n$ that takes $\mathbf{x} \in \mathbb{R}^n$ to $r\mathbf{x}$ transforms a block B to a block with volume $r^n v^n(B)$. Then this is true for finite unions of blocks and, therefore, for the inner and outer approximations of Jordan sets. Then we see that the volume of any Jordan set is multiplied by r^n under this scaling transformation.

Now let $\lambda(y) = (1 - y^2)^{1/2}$ for $-1 < y < 1$. We see that the cross-section of the unit ball $B_r^{n+1}(\mathbf{0}) \subset \mathbb{R}^{n+1} = \mathbb{R}^n \times \mathbb{R}$ with y=constant space is

$$B^n_{\lambda(y)} \text{ if } -1 < y < 1 \text{ and empty otherwise.}$$

For each interval $I \subset (-1, 1)$ let $\alpha(I) = \inf_{t \in I} \lambda(t)$ and $\beta(t) = \sup_{t \in I} \lambda(t)$. Then we see that $V_I(E) \le (\beta(I)^n - \alpha(I)^n) a_n$. This shows that the cross sections are continuously changing. Hence $B_1^{n+1}(\mathbf{0})$ is a Jordan set by Problem 8.10.

8.19 Show that the region

$$E = \{ (x, y, z) \mid x^2 + z^2 \le 1,\ y^2 + z^2 \le 1 \}$$

is a Jordan set in $\mathbb{R}^3$. This the region between two (ordinary, circular) cylinders.

Solution. In problem 8.9, we derived the variation of cross-sections of E with horizontal planes. From the solution, it's clear that if $J = [a, b]$ is an interval on the z-axis, then $V_E(J)$ is a continuous function of a and b. Hence E is a Jordan set, once again by problem 8.10.

8.21 Give an example of a bounded set E in $\mathbb{R}^2$ such that E is not a Jordan set in $\mathbb{R}^2$ but all its cross-sections $E_y \subset \mathbb{R}$ are Jordan sets in $\mathbb{R}$ with $v(E_y) = \ell(E_y) = 1/2$ for all y in the interval $[0, 1]$.

Solution. Let E be the set of all points $(x, y) \in \mathbb{R}^2$ such that either (1) y is rational and $0 \le x \le 1/2$, or (2) y is irrational and $1/2 \le x \le 1$.

Then E is contained in the unit square, and each cross-section E_y is an interval of length $1/2$ (and thus a Jordan set in $\mathbb{R}$). To see that E is not a Jordan set in $\mathbb{R}^2$, note that if A is any cube in $\mathcal{C}_k^n$ lying in the unit square, then A will contain both points in E and points not in E. This shows that the inner area of E is 0 but the outer area is 1; hence, E is not a Jordan set.

8.2 INTEGRALS

8.23 Volumes of cones. Let A be a Jordan set in $\mathbb{R}^n$ and $\mathbf{c} = (\mathbf{a}, h) \in \mathbb{R}^n \times \mathbb{R}$ with $h \neq 0$. Show that the volume of cone

$$K = K(A, \mathbf{c}) = \{ (1 - t)(\mathbf{x}, 0) + t(\mathbf{a}, h) \mid \mathbf{x} \in A, \ 0 \leq t \leq 1 \} \subset \mathbb{R}^{n+1}$$

is $v^{n+1}(K) = |h| \, v^n(A)/(n+1)$. (Cones are Jordan sets by Problem 8.14.)

Solution. Let $W = \mathbb{R}^{n+1} = \mathbb{R}^n \times \mathbb{R}$ and write $\mathbf{w} = (\mathbf{u}, v) \in \mathbb{R}^{n+1}$ with $\mathbf{u} \in \mathbb{R}^n$ and $v \in \mathbb{R}$. The non-empty cross-sections K_v of K with constant $v = th$ are all of the form

$$K_v = \{ (1 - t)\mathbf{x} + t\mathbf{a} \mid \mathbf{x} \in A \},$$

where $0 \leq t = v/h \leq 1$. So

$$K_v = (1 - v/h)A + t\mathbf{a}.$$

Using the fact that volume is invariant under translation and the result about scaling noted in problem 8.17, we have $v^n(K_v) = (1 - v/h)^n \, v^n(A)$. Hence by Fubini's Theorem, if $h > 0$ we have

$$\begin{aligned} v^{n+1}(K) &= \int_0^h v^n(K_v) \, dv \\ &= v^n(A) \int_0^h (1 - v/h)^n \, dv \\ &= h \, v^n(A) \int_0^1 (1 - t)^n \, dt \\ &= |h| \, v^n(A)/(n+1). \end{aligned}$$

Here we have used an ordinary change of variable $v = ht$ to compute the integral. The case $h < 0$ is handled similarly, with adjustment of the limits of integration.

8.25 Find the volume of the ellipsoid

$$E = \left\{ \mathbf{x} = (x_1, \ldots, x_n) \in \mathbb{R}^n \;\middle|\; \sum\nolimits_i r_i^{-2}(x_i - c_i)^2 \leq R^2 \right\}.$$

Here $\mathbf{c} = (c_1, \ldots, c_n)$ is the center of the ellipsoid and $(r_1, \ldots, r_n)$ is a fixed vector in $\mathbb{R}^n$ with $r_i \neq 0$. (Ellipsoids are Jordan sets by Problem 8.18.)

Solution. The volume of $E - \mathbf{c}$ is the same as the volume of E, so we may assume $\mathbf{c} = \mathbf{0}$. Hold fixed $(r_1, \ldots, r_n)$ and write ψ_k for the volume of the ellipsoid

$$E_k = \left\{ \mathbf{x} = (x_1, \ldots, x_k) \in \mathbb{R}^k \;\middle|\; \sum_i r_i^{-2} x_i^2 \leq 1 \right\},$$

for $k = 1 \cdots n$. It is enough to find the values ψ_k, where $R = 1$, because the volume for general R is $R^k\, \psi_k$. Finally, we may assume each $r_i > 0$ since only r_i^2 appears in the definition.

The proof is by induction on the number of dimensions, and is very similar to Example 8.2.44. In fact, if ϑ_n is the volume of the unit ball in $\mathbb{R}^n$ as in that example, then we prove by induction that ψ_n is $|r_1|\,|r_2| \cdots |r_n| \vartheta_n$. Since $\psi_1 = \ell(-r_1, r_1) = 2\,r_1 = r_1\, \vartheta_1$, the result holds for $n = 1$.

Assume for induction that $\psi_n = |r_1|\,|r_2| \cdots |r_n| \vartheta_n$. We compute ψ_{n+1}. If $(\mathbf{x}, v) \in \mathbb{R}^{n+1}$ with $v = x_{n+1} \in \mathbb{R}$, the cross-sections of the ellipsoid in $\mathbb{R}^{n+1}$ with $v =$ constant are of the form

$$E_v = \left\{ \mathbf{x} = (x_1, \ldots, x_n) \;\middle|\; \sum_1^n r_i^{-2} x_i^2 \leq 1 - v^2/{r_{n+1}}^2 \right\}.$$

The cross-sections are non-trivial if $|v| \leq r_{n+1}$, and are themselves ellipsoids in $\mathbb{R}^n$ of radius $\sqrt{1 - v^2/{r_{n+1}}^2}$. So by scaling, $v^n(E_v) = (1 - v^2/{r_{n+1}}^2)^{n/2}\, \psi_n$. This gives

$$\begin{aligned}
\psi_{n+1} &= \int_{-r_{n+1}}^{r_{n+1}} v^n(E_v)\, dv \\
&= 2\psi_n \int_0^{r_{n+1}} (1 - v^2/{r_{n+1}}^2)^{n/2}\, dv \\
&= 2\,|r_1|\,|r_2| \cdots |r_n|\, \vartheta_n |r_{n+1}| \int_0^1 (1 - t^2)^{n/2}\, dt \\
&= |r_1|\,|r_2| \cdots |r_{n+1}|\, \vartheta_{n+1}.
\end{aligned}$$

In the third equality, we have used the inductive assumption and ordinary change of variable $v = t r_{n+1}$. In the fourth equality, we have used the relation between ϑ_{n+1} and ϑ_n derived in Example 8.2.44.

It follows that, in general, $v^n(E) = |r_1| \cdots |r_n| R^n\, \vartheta_n$.

8.27 Let $p \in \mathbb{R}$. Find the volume V of

$$E = \{\, (x, y, z) \mid x^2 + y^2 + z^2 \leq 1,\ p \leq z \,\}.$$

How do we know that E is a Jordan set?

Solution. E is the portion of the unit ball that lies on or above the plane $z = p$. If $p > 1$ then E is empty and has zero volume. If $p \le 1$ then the cross-sections of E change continuously on the interval $[-1, 1] \cap [p, 1]$, so that E is a Jordan set. Assume $-1 \le p \le 1$, since otherwise E is the unit ball in $\mathbb{R}^3$ and $V = 4/3\pi$. Then by Fubini's Theorem we have

$$\begin{aligned} V(E) &= \int_p^1 \pi(1 - z^2)\, dz \\ &= \pi(2/3 - p + p^3/3). \end{aligned}$$

8.29 Give an example of a non-integrable function $f : \mathbb{R}^2 \to \mathbb{R}$ for which the integral $\int \int f(x, y)\, dx\, dy$ exists. (This means that $g(y) = \int f(x, y)\, dx$ exists for each y and defines an integrable function $g : \mathbb{R} \to \mathbb{R}$. Such an example shows that the converse of Fubini's theorem is false.)

Solution. Let $f(x, y) = 1$ if $0 \le y \le 1$ and either (1) y is rational and $0 \le x \le 1/2$, or (2) y is irrational and $1/2 \le x \le 1$. In other words, f is the characteristic function for the set E in our solution to Problem 8.21. We showed that E is not a Jordan set, which implies that f is not integrable. Nevertheless, for each fixed $y \in [-1, 1]$, either y is rational and

$$g(y) = \int f(x, y)\, dx = \int_0^{1/2} 1\, dx = 1/2,$$

or else y is irrational and

$$g(y) = \int f(x, y)\, dx = \int_{1/2}^{1} 1\, dx = 1/2.$$

For $y > 1$ or $y < 1$, the integrals are 0. Thus all relevant integrals $g(y)$ exist. Furthermore, $\int \int f(x, y)\, dx\, dy = \int_0^1 1/2 = 1/2$.

8.31 If $f : \mathbb{R}^2 \to \mathbb{R}$ is integrable, show that for any $\varepsilon > 0$,

$$E_\varepsilon = \left\{ y \in \mathbb{R} \,\middle|\, \overline{\int} f(x, y)\, dx - \underline{\int} f(x, y)\, dx > \varepsilon \right\}$$

is a Jordan subset of $\mathbb{R}$ with $v(E_\varepsilon) = 0$.

Solution. Suppose $\overline{v}(E_\varepsilon) > 0$. Then there is a Jordan set $A \subset E_\varepsilon$ with $v(A) = k$, $k > 0$. Fubini's theorem implies that

$$\int \overline{\int} f(x, y)\, dxdy = \int \underline{\int} f(x, y)\, dxdy.$$

But

$$\int (\overline{\int} f(x,y)\,dx - \underline{\int} f(x,y)\,dx)dy \geq \int_A (\overline{\int} f(x,y)\,dx - \underline{\int} f(x,y)\,dx)dy$$
$$\geq kv(A) > 0,$$

a contradiction.

8.33 Show that if a bounded function of compact support is integrable, then the set E of its discontinuities is a countable union of negligible sets E_i. That is, there is a sequence of negligible sets E_i, $i \in \mathbb{N}$, such that $E = \cup_{i\in\mathbb{N}} E_i$. Give an example to show that the set E of discontinuities need not itself be negligible.

Solution. Suppose $f : \mathbb{R}^n \to \mathbb{R}$ is an integrable function of compact support. By problem 4.97, f is continuous at $\mathbf{a}$ if and only if $\omega(f, \mathbf{a}) = 0$, where ω is the oscillation as defined in problem 4.96. Let $E_i = \{\, \mathbf{z} \mid \omega(f, \mathbf{z}) \geq \frac{1}{i} \,\}$. Then the set of discontinuities of f is the countable union of the sets E_i, and by Problem 8.32, E_i is negligible for all $i \in \mathbb{N}$.

8.35 Let $0 < \beta$. Let K and B_k be compact Jordan sets such that $B_k \subset K$ and $\beta \leq v(B_k)$ for all $k \in \mathbb{N}$. Then show that there is an $\mathbf{x} \in K$ that belongs to infinitely many B_k s.

Solution. 1. Note that $v(\cup_{j=1}^n B_j) \leq v(K)$ for all $n \in \mathbb{N}$. Hence $\sup_n v(\cup_{j=1}^n B_j)$ exists. Given $\varepsilon > 0$ find an $m \in \mathbb{N}$ such that

$$\sup_n v(\cup_{j=1}^n B_j) \leq v(\cup_{j=1}^m B_j) + \varepsilon.$$

Let $K' = \cup_{j=1}^m B_j$. We claim that $v(K' \cap B_k) \geq \beta - \varepsilon$ for all $k \in \mathbb{N}$. In fact $K' \cup B_k$ is the disjoint union of K' and $B_k \setminus K'$. Therefore

$$v(K') + v(B_k \setminus K') = v(K' \cup B_k) \leq \sup_n v(\cup_{j=1}^n B_j) \leq v(K') + \varepsilon.$$

Hence $v(B_k \setminus K') \leq \varepsilon$. Then

$$\beta \leq v(B_k) = v(K' \cap B_k) + v(B_k \setminus K') \leq v(K' \cap B_k) + \varepsilon,$$

since B_k is the disjoint union of $K' \cap B_k$ and $B_k \setminus K'$.

2. We will define a sequence of integers $n_{i-1} < n_i$, $n_0 = 0$, and let

$$K_0 = K, \quad K_i = \cup\{\, K_{i-1} \cap B_j \mid n_{i-1} < j \leq n_i \,\}, \quad i \in \mathbb{N}. \tag{8.1}$$

Each K_i is a compact Jordan set and $K_{i+1} \subset K_i$. If n_i s are such that $v(K_i) \neq 0$ for each $i \in \mathbb{N}$ then Problem 4.77 shows that $L = \cap_{i\in\mathbb{N}} K_i$ is nonempty. In this case any $\mathbf{x} \in L$ belongs to infinitely many B_j. In fact there is at least one j in each one of the segments $n_{i-1} < j \leq n_i$ such that $\mathbf{x} \in B_j$.

Find n_i s by induction. Let $\varepsilon_i > 0$, $i \in \mathbb{N}$, be such that $\sum_i \varepsilon_i \leq (\beta/2)$. At the initial stage we have $n_0 = 0$, a compact Jordan set $K_0 = K$, a sequence of compact Jordan sets $B_j^0 = B_j \subset K_0$, $j \in \mathbb{N}$, such that $v(B_j^0) \geq \beta_0 = \beta - \varepsilon_0$ for all $j \in \mathbb{N}$. To proceed to the next stage we find an $m_1 \in \mathbb{N}$, as in Part 1, such that

$$\varepsilon_1 + v(\cup_{j=1}^{m_1} B_j^0) \geq \sup_m v(\cup_{j=1}^{m} B_j^0).$$

We then let $n_1 = n_0 + m_1$ and $K_1 = \cup_{j=1}^{n_1} B_j^0$. This agrees with the definition of a general K_i given in (8.1). Now let

$$B_j^1 = K_1 \cap B_{n_1+j},\ j \in \mathbb{N}.$$

The argument given in Part 1 shows that $v(B_j^1) \geq \beta_1 = (\beta_0 - \varepsilon_1)$ for all $j \in \mathbb{N}$. At a general step $i \in \mathbb{N}$ assume that n_i and the compact Jordan sets K_i and $B_j^i \subset K_i$, $j \in \mathbb{N}$, are obtained and that they satisfy

$$v(B_j^i) \geq \beta_i = \beta - (\varepsilon_1 + \cdots + \varepsilon_i) > \beta/2 \tag{8.2}$$

for all $j \in \mathbb{N}$. Now find an $m_{i+1} \in \mathbb{N}$ such that

$$\varepsilon_{i+1} + v(\cup_{j=1}^{m_{i+1}} B_j^i) \geq \sup_m v(\cup_{j=1}^{m} B_j^i).$$

We then let $n_{i+1} = n_i + m_{i+1}$ and $K_{i+1} = \cup_{j=1}^{m_{i+1}} B_j^i$. Again note that this agrees with the definition of a general K_i given in (8.1). Now let

$$B_j^{i+1} = K_{i+1} \cap B_{n_{i+1}+j},\ j \in \mathbb{N}.$$

The argument in Part 1 shows that (8.2) still holds for $i + 1$. Hence $v(K_i) \neq 0$ for all $i \in \mathbb{N}$ since $v(K_i) \geq v(B_j^{i-1}) \geq \beta/2$.

8.37 Let f_n be a sequence of integrable functions. Assume that all f_n s have support in a compact Jordan set K and that $|f_n(\mathbf{x})| \leq M$ for all $n \in \mathbb{N}$ and for all $\mathbf{x} \in K$. If $\lim_n f_n(\mathbf{x}) = f(\mathbf{x})$ for all $\mathbf{x} \in K$ and if f is also integrable, then show that $\lim_n \int f_n = \int f$.

Solution. Let $g_n = |f_n - f|$. Then g_ns are integrable functions, $0 \leq g_n(\mathbf{x}) \leq 2M$, and $\lim_n g_n(\mathbf{x}) = 0$, for all $\mathbf{x} \in K$. We will show that $\lim_n \int g_n = 0$.

In fact, if this is not the case then there is a $\delta > 0$ such that $\delta \leq \int g_n$ for infinitely many n s. Assume $\delta \leq \int g_n$ for all $n \in \mathbb{N}$, without loss of generality. Given $\alpha > 0$ let D_n be the set of $\mathbf{x} \in K$ such that $\alpha \leq g_n(\mathbf{x})$. Then $\delta \leq 2Mv(D_n) + \alpha v(K)$ or $(1/2M)(\delta - \alpha v(K)) \leq v(D_n)$. This follows from a simple direct argument or from Problem 8.36. Choose $\alpha > 0$ sufficiently small so that $(1/M)(\delta - \alpha v(K)) = 2\beta > 0$. Then each D_n contains a compact Jordan set $B_n \subset K$ with $0 < \beta \leq v(B_n)$. In this case Problem 8.35 shows that there are $\mathbf{x} \in K$ that belongs to infinitely many

B_n s. This means that $g_n(\mathbf{x})$ can not converge to 0. This contradiction shows that $\lim_n \int g_n = 0$. Then the solution follows from $|\int f_n - \int f| \le \int |f_n - f|$.

8.39 Give a counterexample to show that the conclusion in Problem 8.37 is false if there is no number M which is an upper bound for all $|f_n|$s.

Solution. Let $K = [0, 1]$ and let $f_n : \mathbb{R} \to \mathbb{R}$ be a function such that $f_n(x) = 0$ if $x \le 0$ or $x \ge 1$ or $0 \le x \le 1 - (1/2^n)$. On the interval $[1 - (1/2^n), 1]$, the graph of f_n is an isosceles triangle with top vertex at $(1 - 1/2^{n+1}, 2^n)$. Thus, $\int f_n = 1/2$ for each n. Even though $\lim_n f_n(x) = f(x) = 0$ for all x in $[0, 1]$, and $f(x) = 0$ is integrable, we have $\lim_n \int f_n = 1/2 \ne 0 = \int f$.

8.41 Give examples of f_n and f to show that several of the hypotheses in Problem 8.37 are not necessary for the conclusion of the problem to be true.

Solution. a) The assumption of an upper bound M is not necessary. Modify the example in the solution for 8.39: in defining f_n, let the top vertex of the triangle be $(1 - 1/2^{n+1}, n)$. We still have $\lim_n f_n(x) = f(x) = 0$, but now $\int f_n = (1/2)n/2^n$. This converges to $0 = \int f$.

b) The assumption of a compact Jordan set K containing the support of all f_n is not necessary. Let $f_n : \mathbb{R} \to \mathbb{R}$ be defined by $f_n(x) = 1/n$ if $n \le x < n + 1$ and $f_n(x) = 0$ otherwise. Clearly, $\lim_n f_n(x) = f(x) = 0$, and also

$$\lim_n \int f_n = \lim_n (1/n) = 0 = \int f.$$

8.3 IMAGES OF JORDAN SETS

8.43 Find the volume of $E = \{ (x, y) \mid (2x + y)^2 + (x - y)^2 \le 1 \} \subset \mathbb{R}^2$.

Solution. Put $T(x, y) = (2x + y, x - y)$. Then T is an isomorphism of $\mathbb{R}^2$. If B is the unit circle, then $T(x, y) \in B$ if and only if $(x, y) \in E$, and therefore $T(E) = B$. So $E = T^{-1}(B)$ and by Theorem 8.3.16, $v(E) = (1/|\det T|)\, v(B)$. A simple calculation gives $\det T = -3$. So $v(E) = \pi/3$.

8.45 Find the volume of

$$E = \{ (x, y, z) \mid |x| + |x + y| + |x + y + z| \le 1 \} \subset \mathbb{R}^3.$$

Solution. First, the volume of the tetrahedron $A = \{ (x, y, z) \mid |x| + |y| + |z| \le 1 \}$ in $\mathbb{R}^3$ can be computed using cross-sections:

$$\begin{aligned} v(A) &= \int_{-1}^{1} 2(1 - |z|)^2 \, dz \\ &= 4/3. \end{aligned}$$

Now put $T(x, y, z) = (x, x + y, x + y + z)$. Then T is an isomorphism of $\mathbb{R}^3$ and $T(E) = A$. Hence $v(E) = (1/|\det T|)\, v(A)$. Since $\det T = 1$, we have $v(E) = 4/3$.

8.47 Show that any open ball or any closed ball with respect to any norm on $\mathbb{R}^n$ is a non-negligible Jordan set.

Solution. The result follows directly from Problem 8.46. If $||\cdot||$ is any norm on $\mathbb{R}^n$, $H_r = \{\, \mathbf{x} \mid ||\mathbf{x}|| < r \,\} \subset \mathbb{R}^n$ and $A_r = \{\, \mathbf{x} \mid ||\mathbf{x}|| \le r \,\} \subset \mathbb{R}^n$, then we know that H_r is open, A_r is compact, and $tA_r \subset H_r$ whenever $0 < t < 1$. So both H_r and A_r are Jordan sets. Also, by the equivalence of norms, the sets H_r and A_r contain the usual unit ball of $\mathbb{R}^n$ if r is sufficiently large, and thus these sets have non-negligible volume. By scaling, H_1 and A_1 are non-negligible and hence H_r and A_r are non-negligible whenever $r > 0$.

We can also show that H_r and A_r are Jordan sets without appealing to Problem 8.46. We know that H_r is bounded. Let $G = O_k(H_r)$ be the kth order outer approximation of H_r, a union of cubes in $\mathcal{C}_k^n$. By equivalence of norms, we can choose k so that the radius (in the chosen norm) of the cubes is less than ε, for any desired $\varepsilon > 0$. So G lies entirely in $A_{r+\varepsilon}$, and if $t = r/(r+\varepsilon)$, then $F = tG$ is a union of cubes and $F \subset H_r \subset A_r \subset G$. Then $v(G) - v(F) \le (1 - t^n)M$, where $M > 0$ is an upper bound on $v(G)$. Since t^n approaches 1 as ε approaches 0, this proves that H_r and A_r are Jordan sets. (Note that this is essentially the proof of 8.46.)

8.49 Denote the vectors in $\mathbb{R}^{n+1} = \mathbb{R}^n \times \mathbb{R}$ as $(\mathbf{x}, y)$ with $\mathbf{x} \in \mathbb{R}^n$ and $y \in \mathbb{R}$. Let $\mathbf{a} \in \mathbb{R}^n$ be a fixed vector. Define $R : \mathbb{R}^{n+1} \to \mathbb{R}^{n+1}$ by $R(\mathbf{x}, y) = (\mathbf{x} + y\mathbf{a}, y)$ for all $(\mathbf{x}, y) \in \mathbb{R}^{n+1}$. Then show that $\rho(R) = 1$.

Solution. First, R is an isomorphism of $\mathbb{R}^{n+1}$. Let $(\mathbf{e}_1, \ldots, \mathbf{e}_{n+1})$ be the usual orthonormal basis of $\mathbb{R}^{n+1}$. Then $R(\mathbf{e}_i) = \mathbf{e}_i$ for $1 \le i \le n$, and $R(\mathbf{e}_{n+1}) = a_1\mathbf{e}_1 + \cdots + a_n\mathbf{e}_n + \mathbf{e}_{n+1}$, where $a = a_1\mathbf{e}_1 + \cdots + a_n\mathbf{e}_n$. From the matrix representation of R, it is clear that $\det R = 1$, so that $\rho(R) = 1$.

8.51 Show that any bounded convex set is a Jordan set.

Solution. Let E be a bounded convex set in $\mathbb{R}^n$. By problem 4.92, either E is contained in a lower-dimensional subspace or E contains an interior point. If E is contained in a lower-dimensional subspace, then obviously E is a Jordan set (since it is bounded) and $v(E) = 0$.

If E has an interior point $\mathbf{a}$, then note that E is Jordan just in case the translated set $E - \mathbf{a}$ is Jordan, and E is convex just in case $E - \mathbf{a}$ is convex. So we can assume that $\mathbf{0}$ is an interior point. Then $H = E^o$ is open and $A = \overline{E}$ is compact, and $H \subset E \subset A$, as in the statement of problem 8.46. Also, by problem 4.94, if $\mathbf{b} \in A = \overline{E}$ and $0 \le t < 1$, then $t\mathbf{b} \in H = E^o$. It follows from problem 8.46 that E is a Jordan set.

8.53 Let E be an open set in $X = \mathbb{R}^m$ and F an open set in $Y = \mathbb{R}^n$. Let $\mathbf{u} : E \to U = \mathbb{R}^m$ and $\mathbf{v} : F \to V = \mathbb{R}^n$ be two diffeomorphisms. Show that $\mathbf{w}(\mathbf{x},\, \mathbf{y}) = (\mathbf{u}(\mathbf{x}),\, \mathbf{v}(\mathbf{y}))$ defines a diffeomorphism

$$\mathbf{w} : (E \times F) \to W = \mathbb{R}^{m+n} = \mathbb{R}^m \times \mathbb{R}^n.$$

Also show that $\rho(\mathbf{w}'(\mathbf{x},\, \mathbf{y})) = \rho(\mathbf{u}'(\mathbf{x}))\, \rho(\mathbf{v}'(\mathbf{y}))$ for $\mathbf{x} \in E$, $\mathbf{y} \in F$.

Solution. First, $G = \mathbf{u}(E)$ and $H = \mathbf{v}(F)$ are open sets, so that $G \times H$ is open. It is trivial to show that $\mathbf{w}$ is a one-to-one and onto mapping between $E \times F$ and $G \times H$.

Set

$$\begin{aligned} P(\mathbf{x}, \mathbf{y}) &= \mathbf{x} \\ Q(\mathbf{x}, \mathbf{y}) &= \mathbf{y} \\ R\mathbf{x} &= (\mathbf{x}, \mathbf{0}) \\ S\mathbf{y} &= (\mathbf{0}, \mathbf{y}); \end{aligned}$$

all of these are clearly linear functions and hence $\mathcal{C}^1$. We see that

$$\mathbf{w} = R\mathbf{u}P + S\mathbf{v}Q.$$

Since $\mathbf{u}$ and $\mathbf{v}$ are $\mathcal{C}^1$, it follows that $\mathbf{w}$ is $\mathcal{C}^1$, by the Chain Rule. Using the Chain Rule and the fact that the derivative of a linear function is constant, for any $(\mathbf{a},\, \mathbf{b}) \in E \times F$ we have

$$\mathbf{w}'(\mathbf{a},\, \mathbf{b}) = R\mathbf{u}'(\mathbf{a})P + S\mathbf{v}'(\mathbf{b})Q,$$

so that

$$\mathbf{w}'(\mathbf{a},\, \mathbf{b})(\mathbf{x},\, \mathbf{y}) = (\mathbf{u}'(\mathbf{a})\mathbf{x},\, \mathbf{v}'(\mathbf{b})\mathbf{y}).$$

It is clear that the kernel of $\mathbf{w}'(\mathbf{a},\, \mathbf{b})$ is $(\mathbf{0},\, \mathbf{0})$ and hence it is an invertible linear transformation. By the inverse function theorem, it follows that the inverse of $\mathbf{w}$ is $\mathcal{C}^1$. Therefore $\mathbf{w}$ is a diffeomorphism.

For the volume multiplier, let $\mathbb{A}$ be an eigenbasis for $\mathbf{u}'(\mathbf{x})$ and let $\mathbb{B}$ be an eigenbasis for $\mathbf{v}'(\mathbf{y})$. Then $R\mathbb{A} \cup S\mathbb{B}$ is an eigenbasis for $\mathbf{w}$ and we can mimic the proof of Theorem 8.3.17 to see that

$$\det \mathbf{w}'(\mathbf{x},\, \mathbf{y}) = \det \mathbf{u}'(\mathbf{x}) \cdot \det \mathbf{v}'(\mathbf{y}).$$

8.4 CHANGE OF VARIABLES

8.55 Integrate $f(x,\, y) = 2x^2 + y^2$ over the region in the first quadrant (that is $x \geq 0$ and $y \geq 0$) bounded by the curves $xy = 2$, $xy = 4$, $y = 3x$, and $y = 5x$.

Solution. Put $u = xy$ and $v = y/x$ and let $\varphi(x, y) = (u, v)$. If F is the region described, then $\varphi(F) = E = [2, 4] \times [3, 5]$. The inverse function $\varphi^{-1} : E \to F$ is given by $x = \sqrt{\frac{u}{v}})$ and $y = \sqrt{uv}$. The Jacobian determinant is easily computed as $\det(\varphi'(x, y)) = \frac{2y}{x}$, so that the Jacobian determinant for the inverse function is $\det(\varphi^{-1})'(u, v) = \frac{1}{2v}$. By the change of variables theorem, we have

$$\begin{aligned} \int_F (2x^2 + y^2)\, dx\, dy &= \int_{\varphi^{-1}(E)} (2(u/v) + uv)(1/2v)\, du\, dv \\ &= \int_3^5 \int_2^4 \left(\frac{u}{v^2} + \frac{u}{2}\right) du\, dv \\ &= \int_3^5 \left(\frac{6}{v^2} + 3\right) dv \\ &= \frac{34}{5}. \end{aligned}$$

8.57 The elliptical region $(x/a)^2 + (y/b)^2 \leq 1$ is divided into two parts by the ellipse $(x/a)^2 + ((y-b)/b)^2 = 1$ into two parts. Find the areas of these parts.

Solution. First consider the case of the unit circle B, where $a = b = 1$. The area of the region of intersection E can be computed by Fubini's theorem:

$$\begin{aligned} \text{Area}(E) &= \int_0^{1/2} 2\sqrt{1-(y-1)^2}dy + \int_{1/2}^1 2\sqrt{1-y^2}dy \\ &= 4\int_{1/2}^1 \sqrt{1-y^2}dy \text{ by ordinary change of variable} \\ &= 2\left(y\sqrt{1-y^2} + \arcsin y\right)\Big|_{1/2}^1 \\ &= (2\pi/3) - (\sqrt{3}/2). \end{aligned}$$

Hence the remaining region of the unit circle, $B \setminus E$, has area $\pi/3 + \sqrt{3}/2$.

For the general ellipse, we want the areas of TE and $T(B \setminus E)$, where $T(x, y) = (ax, by)$. Since $\det T = ab$, we just multiply the above areas for E and $B \setminus E$ by ab.

8.59 Let $f(x, y) = \exp(-x^2 - y^2)$. Compute the integral of f over a disc $x^2 + y^2 \leq R^2$.

Solution. Using polar coordinates $x = rcos\theta$, $y = rsin\theta$, the Jacobian determinant is r. Also, the disc D_R is the image of $[0, R] \times [0, 2\pi]$. So we have

$$\begin{aligned}\int_{D_R} \exp(-x^2 - y^2)\, dxdy &= \int_0^{2\pi} \int_0^R \exp(-r^2) r\, drd\theta \\ &= (2\pi)\left(-\frac{1}{2}\exp(-r^2)\Big|_0^R\right) \\ &= \pi(1 - \exp(-R^2)).\end{aligned}$$

8.61 Let $T : \mathbb{R}^2 \to \mathbb{R}^2$ be an isomorphism. For each R let E_R be the elliptical region $(x/a)^2 + (y/b)^2 \leq R^2$. Here $a \neq 0$ and $b \neq 0$ are fixed. Show that $\lim_{R\to\infty} \int_{E_R} \exp(-\langle T\mathbf{z}, T\mathbf{z}\rangle)\, d\mathbf{z}$ exists and compute its value.

Solution. We use the change of variable $\mathbf{x} = T\mathbf{z}$. We have

$$\int_{E_R} \exp(-\langle T\mathbf{z}, T\mathbf{z}\rangle)\, d\mathbf{z} = (1/|detT|) \int_{TE_R} \exp(-x^2 - y^2)\, dx\, dy,$$

where we have put $\mathbf{x} = (x, y)$. Since T is an isomorphism, the mapping $\mathbf{z} \to \|T\mathbf{z}\|$ is a norm, and hence equivalent to the Euclidean norm. So there are α and β such that $\alpha\|\mathbf{z}\| \leq \|T\mathbf{z}\| \leq \beta\|\mathbf{z}\|$ for all $\mathbf{z}$. If $a < b$, then $D_{a\alpha R} \subset TE_R \subset D_{b\beta R}$; if $b \leq a$, then $D_{b\alpha R} \subset TE_R \subset D_{a\beta R}$. Either way, TE_R is sandwiched between two disks of the form D_{cR}. By problem 8.60, the limit as $R \to \infty$ of the integral of $\exp(-x^2 - y^2)$ over any disk D_{cR} is π. Hence,

$$\lim_{R\to\infty} \int_{E_R} \exp(-\langle T\mathbf{z}, T\mathbf{z}\rangle)\, d\mathbf{z} = \frac{\pi}{|\det T|}.$$

8.63 Centroid. Let E be a Jordan set in $\mathbb{R}^n$. Show that there is a unique vector $\mathbf{c} \in \mathbb{R}^n$, called the *centroid* of E, such that

$$\langle \mathbf{a}, \mathbf{c}\rangle = \frac{1}{v(E)} \int_E \langle \mathbf{a}, \mathbf{z}\rangle\, d\mathbf{z}$$

for all $\mathbf{a} \in \mathbb{R}^n$.

Solution. For any $\mathbf{a}$, the function $h(\mathbf{z}) = \langle \mathbf{a}, \mathbf{z}\rangle$ is linear; therefore h is continuous and integrable. Now suppose E is a fixed Jordan set in $\mathbb{R}^n$ with $v(E) \neq 0$. Let

$$f(\mathbf{a}) = \frac{1}{v(E)} \int_E \langle \mathbf{a}, \mathbf{z}\rangle\, d\mathbf{z}.$$

It is easy to see that $f : \mathbb{R}^n \to \mathbb{R}$ is a linear transformation. Hence by Theorem 3.4.21 on the representation of linear functionals, there exists a unique vector $\mathbf{c} \in \mathbb{R}^n$ such that

$$\langle \mathbf{a}, \mathbf{c}\rangle = f(\mathbf{a}) = \frac{1}{v(E)} \int_E \langle \mathbf{a}, \mathbf{z}\rangle\, d\mathbf{z}$$

for all $\mathbf{a} \in \mathbb{R}^n$.

8.65 Find the volume of the torus obtained by rotating the disc $(x-2)^2 + z^2 \leq 1$ around the z-axis.

Solution. The radius of the disc is 1, so its area is π. If we can assume the centroid of the disc is $\mathbf{c} = (2, 0)$, then by Pappus' Theorem (problem 8.64), the volume of the torus is $(2\pi)(2\pi) = 4\pi^2$. So the only thing to show is that the centroid is as claimed. To prove this, it suffices (by linearity) to show that the equality in problem 8.63 holds with $\mathbf{c} = (2, 0)$ if $\mathbf{a} = (1, 0)$ or $\mathbf{a} = (0, 1)$. Both of these can be established by an easy integration.

CHAPTER 9

INTEGRATION ON MANIFOLDS

9.1 EUCLIDEAN VOLUMES

9.1 Let (X, Y) and (U, V) be two orthogonal coordinate systems (Definition 3.1.42) in Z. Assume that (U, Y) is also a coordinate system for Z. Let $P : U \to X$ be the orthogonal projection of U on X and let $Q : Y \to V$ be the orthogonal projection of Y on V. Show that $\rho(P) = \rho(Q)$.

Solution. This problem is essentially the same as Problem 3.92. It can be solved by an application of Theorem 3.6.20. Here is another solution. Let $\mathbb{U} = (\mathbf{u}_1, \ldots, \mathbf{u}_k)$ be a basis for U and let $\mathbb{Y}$ be a basis for Y. Then

$$\vartheta_Z(\mathbb{U}; \mathbb{Y}) = \vartheta_Z(P\mathbb{U}; \mathbb{Y}). \tag{9.1}$$

In fact, if $\mathbf{u}_i = P\mathbf{u}_i + \mathbf{v}_i$ then $\mathbf{v}_i \in Y$. Hence

$$\begin{aligned} \vartheta_Z(\mathbf{u}_1, \ldots, \mathbf{u}_k; \mathbb{Y}) &= \vartheta_Z(P\mathbf{u}_1 + \mathbf{v}_1, \ldots, \mathbf{u}_k; \mathbb{Y}) \\ &= \vartheta_Z(P\mathbf{u}_1, \ldots, \mathbf{u}_k; \mathbb{Y}), \end{aligned}$$

Analysis in Vector Spaces.
By M. A. Akcoglu, P. F. A. Bartha and D. M. Ha

since $\vartheta_Z(\mathbf{v}_1, \ldots, \mathbf{u}_k; \mathbb{Y}) = 0$. Repeating this argument for the other $\mathbf{u}_i$ we obtain (9.1). Similarly, $\vartheta_Z(\mathbb{U}; \mathbb{Y}) = \vartheta_Z(\mathbb{U}; Q\mathbb{Y})$. Now Theorem 9.1.11 shows that

$$\begin{aligned}\vartheta_Z(\mathbb{U}; \mathbb{Y}) &= \gamma\vartheta_U(\mathbb{U})\vartheta_Y(\mathbb{Y}). \text{ Also,}\\ \vartheta_Z(P\mathbb{U}; \mathbb{Y}) &= \pm\vartheta_X(P\mathbb{U})\vartheta_Y(\mathbb{Y}),\end{aligned}$$

by the same theorem, since $X \perp Y$. But $\vartheta_X(P\mathbb{U}) = \pm\rho(P)\vartheta_U(\mathbb{U})$. Hence we see that $\rho(P) = |\gamma|$. By similar arguments, $\rho(Q) = |\gamma|$. Hence $\rho(P) = \rho(Q)$.

9.2 INTEGRATION ON MANIFOLDS

9.3 Find the surface area of a sphere of radius R.

Solution. Let $D = \{ (x, y) \mid x^2 + y^2 < R^2 \}$ be the open disc of radius R in the xy-plane. The upper half the sphere is the graph of $f(x, y) = (R^2 - x^2 - y^2)^{1/2}$, $(x, y) \in D$. Both D and f depend on the combination $x^2 + y^2$ only. Hence we use the polar coordinates to obtain a parametric representation for the upper half of the sphere of the form $\varphi : E \to \mathbb{R}^3$ where

$$\begin{aligned}E &= \{ (r, \vartheta) \mid 0 < r < R,\ 0 < \vartheta < 2\pi \} \text{ and}\\ \varphi(r, \vartheta) &= (r\cos\vartheta,\ r\sin\vartheta,\ (R^2 - r^2)^{1/2}).\end{aligned}$$

Let $\mathbf{e}_1$ and $\mathbf{e}_2$ be the standard basis of the $r\vartheta$-plane. Then

$$\begin{aligned}\varphi'(r, \vartheta)\mathbf{e}_1 &= (\partial\varphi/\partial r)(r, \vartheta) = (\cos\vartheta,\ \sin\vartheta,\ -r(R^2 - r^2)^{-1/2}) \text{ and}\\ \varphi'(r, \vartheta)\mathbf{e}_2 &= (\partial\varphi/\partial\vartheta)(r, \vartheta) = (-r\sin\vartheta,\ r\cos\vartheta,\ 0).\end{aligned}$$

To find the volume multiplier $\rho(\varphi'(r, \vartheta))$ we compute

$$\begin{aligned}\det\{\langle\varphi'(r, \vartheta)\mathbf{e}_i, \varphi'(r, \vartheta)\mathbf{e}_j\rangle\} &= \begin{vmatrix} 1 + r^2(R^2 - r^2)^{-1} & 0 \\ 0 & r^2 \end{vmatrix}\\ &= R^2r^2(R^2 - r^2)^{-1}.\end{aligned}$$

Hence $\rho(\varphi'(r, \vartheta)) = Rr(R^2 - r^2)^{-1/2}$. Therefore, if F is a compact subset of E then the surface area of $\varphi(F)$ is given as

$$\sigma(\varphi(F)) = \int_E Rr(R^2 - r^2)^{-1/2}\, dr\, d\vartheta.$$

The upper half of the sphere is $\varphi(E)$. Note that E is not a compact set. Also, the function to be integrated, $Rr(R^2 - r^2)^{-1/2}$ can not be extended continuously to the closure of E. Hence, instead of E we consider, with sufficiently small $\varepsilon > 0$,

$$E_\varepsilon = [\varepsilon,\ R - \varepsilon] \times [\varepsilon,\ 2\pi - \varepsilon],$$

and obtain

$$\begin{aligned}\sigma(\varphi(E_\varepsilon)) &= \int_{E_\varepsilon} Rr(R^2-r^2)^{-1/2}\,dr\,d\vartheta \\ &= \int_\varepsilon^{R-\varepsilon}\int_\varepsilon^{2\pi-\varepsilon} Rr(R^2-r^2)^{-1/2}\,d\vartheta\,dr \\ &= R(2\pi-2\varepsilon)((R^2-\varepsilon^2)^{1/2}-(2R\varepsilon-\varepsilon^2)^{1/2}).\end{aligned}$$

We see that this converges to $2\pi R^2$ as $\varepsilon \to 0^+$. This is the surface area of half of the sphere. Hence the total surface area of a sphere of radius R is $4\pi R^2$.

9.5 Let E be a Jordan set in the rectangle $(-\pi,\,\pi)\times(0,\,\pi)$. Map this region by the spherical coordinates to get a region $\Phi(E)$ in the unit sphere. Hence $\Phi(E)$ consists of all points with the Cartesian coordinates

$$x=\cos\theta\,\sin\varphi,\quad y=\sin\theta\,\sin\varphi,\quad z=\cos\varphi$$

with $(\theta,\,\varphi)\in E$. Express the surface area of $\Phi(E)$ as an integral over E. Find the areas corresponding to rectangles $-\pi<p\le\theta\le q<\pi, 0<r\le\varphi\le s<\pi$. Also apply this result to give another solution of Problem 9.3.

Solution. Let $\mathbf{e}_1$ and $\mathbf{e}_2$ be the standard basis of the $\theta\varphi$-plane. The parametric representation $\Phi: E\to\mathbb{R}^3$ of the sphere of radius R about the origin is given as

$$\Phi(\theta,\,\varphi)=(R\cos\theta\,\sin\varphi,\;y=R\sin\theta\,\sin\varphi,\;z=R\cos\varphi).$$

To find the volume multiplier $\rho(\Phi'(\theta,\,\varphi))$ we compute

$$\begin{aligned}\det\{\langle\Phi'(\theta,\,\varphi)\mathbf{e}_i,\,\Phi'(\theta,\,\varphi)\mathbf{e}_j\rangle\} &= \begin{vmatrix} R^2\sin^2\varphi & 0\\ 0 & R^2\end{vmatrix} \\ &= R^4\sin^2\varphi.\end{aligned}$$

Note that $0<\sin\varphi$ for $0<\varphi<\pi$. Therefore $\rho(\Phi'(\theta,\,\varphi))=R^2\sin\varphi$. Hence

$$\sigma(\Phi(F))=\int_F R^2\sin\varphi\,d\theta\,d\varphi$$

for compact Jordan subsets F of E. In particular, if F is the rectangle

$$-\pi<p\le\theta\le q<\pi,\;0<r\le\varphi\le s<\pi,$$

then $\sigma(\Phi(F))=R^2(q-p)(\cos r-\cos s)$. If $p\to-\pi^+$, $q\to\pi^-$, $r\to 0^+$, and $s\to\pi^-$, then $\sigma(\Phi(F))\to 4\pi R^2$. This is the surface area of a sphere of radius R.

9.7 Compute the surface area of the helicoidal surface

$$x=r\cos\theta,\quad y=r\sin\theta,\quad z=\theta,$$

where $1 \leq r \leq 2, 0 \leq \theta \leq 2\pi$.

Solution. Define $\varphi : (0, \infty) \times \mathbb{R} \to \mathbb{R}^3$ as

$$\varphi(r, \theta) = (r\cos\theta, r\sin\theta, \theta).$$

Then $\rho(\varphi'(r, \theta))$ is computed as in Problems 9.5 and 9.7 above. We have

$$\begin{aligned}\det\{\langle \Phi'(\theta, \varphi)\mathbf{e}_i, \Phi'(\theta, \varphi)\mathbf{e}_j\rangle\} &= \begin{vmatrix} 1 & 0 \\ 0 & 1+r^2 \end{vmatrix} \\ &= (1+r^2).\end{aligned}$$

Hence the surface area asked in this problem is

$$S = \int_1^2 \int_0^{2\pi} (1+r^2)^{1/2} d\theta\, dr = 2\pi \int_1^2 (1+r^2)^{1/2}\, dr.$$

It should be perfectly acceptable to leave the answer in this form. If necessary, the last integral above can be computed easily to any given degree of accuracy. In this case there is another form of the answer that also refers to outside sources (calculators) for an approximate answer. It is given as

$$S = (\pi/4)(A^2 - A^{-2} - B^2 + B^{-2}) + \pi \log(A/B),$$

where $A = (2 + \sqrt{5})$ and $B = (1 + \sqrt{2})$. To obtain this form let $2r = (e^u - e^{-u})$ and work your way through.

9.9 Find the surface area of the part of the cylinder $x^2 + z^2 = a^2$ that lies above the xy-plane and inside the cylinder $x^2 + y^2 = a^2$.

Solution. A parametric representation for the surface in the problem is

$$\varphi(x, y) = (x, y, (a^2 - x^2)^{1/2}), \;\; u^2 + v^2 < a^2.$$

Proceeding as in the last three problems we obtain

$$\begin{aligned}\det\{\langle \varphi'(x, y)\mathbf{e}_i, \Phi'(x, y)\mathbf{e}_j\rangle\} &= \begin{vmatrix} 1 + x^2(a^2 - x^2)^{-1} & 0 \\ 0 & 1 \end{vmatrix} \\ &= a^2(a^2 - x^2)^{-1}.\end{aligned}$$

Hence $\rho(\varphi'(x, y) = a(a^2 - x^2)^{-1/2}$. Therefore the area asked in the problem is

$$\begin{aligned}\int_{x^2+y^2<1} a(a^2 - x^2)^{-1/2}\, dx\, dy &= \int_{-a}^{a} \int_{-(a^2-x^2)^{1/2}}^{(a^2-x^2)^{1/2}} a(a^2 - x^2)^{-1/2}\, dy\, dx \\ &= \int_{-a}^{a} 2a dx = 4a^2.\end{aligned}$$

9.11 Compute $\int_G (x^2 z + y^2 z) d\sigma$ where G is the upper half of the sphere

$$x^2 + y^2 + z^2 = 4.$$

Solution. We will use the spherical coordinates. The upper half of the sphere of radius $R = 2$ is represented as

$$x = 2\sin\varphi\cos\theta, \quad y = 2\sin\varphi\sin\theta, \quad z = 2\cos\varphi,$$

where $-\pi < \theta < \pi$, $0 < \varphi < \pi/2$. The corresponding volume multiplier is $R^2 \sin\varphi = 4\sin\varphi$, as obtained in Problem 9.5. Also,

$$x^2 z + y^2 z = 8\sin^2\varphi\cos\varphi.$$

Hence the required integral is

$$32 \int_{-\pi}^{\pi} \int_0^{\pi/2} \sin^3\varphi\cos\varphi\, d\varphi\, d\theta = 16\pi.$$

Actually, one has to take the integral first over the compact set

$$[-\pi + \varepsilon, \pi - \varepsilon] \times [\varepsilon, (\pi/2) - \varepsilon]$$

and then let $\varepsilon \to 0^+$. Details are omitted, as they are the same as in Problem 9.3.

9.13 Compute $\int_G (y^2 + z^2)\, d\sigma$ where G is the part of the surface

$$x^2 = 4 - y^2 - z^2$$

that lies in the region $x \geq 0$.

Solution. The region of integration is again a hemisphere. The parametric representation is essentially as in Problem 9.11. We have, with different names for the Cartesian coordinates,

$$y = 2\sin\varphi\cos\theta, \quad z = 2\sin\varphi\sin\theta, \quad x = 2\cos\varphi,$$

where $0 < \theta < 2\pi$, $0 < \varphi < \pi/2$. The volume multiplier is $4\sin\varphi$, as obtained in Problem 9.5. Hence the required integral is

$$\int_0^{2\pi} \int_0^{\pi/2} 4\sin^2\varphi\, 4\sin\varphi\, d\varphi\, d\theta = (64/3)\pi.$$

9.15 Compute $\int_C xy^4\, d\sigma$ where C is the right half of the circle

$$x^2 + y^2 = 16.$$

Solution. A parametric representation for the right half of this circle is $x = 4\cos\theta$, $y = 4\sin\theta$, $-\pi/2 < \theta < \pi/2$. We see that the volume multiplier is

$$\|(x',\, y')\| = \|4(-\sin\theta,\, \cos\theta)\| = 4.$$

Hence the required integral is

$$\int_{-\pi/2}^{\pi/2} (4\cos\theta)(4^4 \sin^4\theta)\, d\theta = 2^{11}/5.$$

9.17 Compute $\int_C xyz\, d\sigma$ where C is the curve

$$x = \sin 2t,\ y = 3t,\ z = \cos 2t,\ 0 \le t \le \pi/4.$$

Solution. The volume multiplier is

$$\|(x',\, y',\, z')\| = \|(2\cos 2t,\ 3,\ -2\sin 2t)\| = \sqrt{13}.$$

Hence

$$\int_C xyz d\sigma = \sqrt{13} \int_0^{\pi/4} 3t \sin 2t \cos 2t\, dt = 3\sqrt{13}\,\pi/32.$$

9.19 Find the centroid (Problem 9.18) of the upper-half of the sphere

$$x^2 + y^2 + z^2 = 1.$$

Solution. Use the spherical coordinates, as in Problem 9.5. The upper half of the sphere $x^2 + y^2 + z^2 = 1$ corresponds to $0 < \theta < 2\pi$ and $0 < \varphi < \pi/2$. The volume multiplier is $R^2 \sin\varphi = \sin\varphi$. The surface area of the half sphere is 2π, as obtained in Problem 9.3. Hence the coordinates of the centroid satisfy

$$\begin{aligned}
2\pi\, x_0 &= \int x\, d\sigma = \int_0^{\pi/2}\int_0^{2\pi} \cos\theta \sin\varphi \sin\varphi\, d\theta\, d\varphi = 0,\\
2\pi\, y_0 &= \int y\, d\sigma = \int_0^{\pi/2}\int_0^{2\pi} \sin\theta \sin\varphi \sin\varphi\, d\theta\, d\varphi = 0,\\
2\pi\, z_0 &= \int z\, d\sigma = \int_0^{\pi/2}\int_0^{2\pi} \cos\varphi \sin\varphi\, d\theta\, d\varphi = \pi.
\end{aligned}$$

Therefore the centroid is the point $(0,\, 0,\, 1/2)$.

9.21 Find the centroid (Problem 9.18) of the helicoidal surface

$$x = r\cos t,\ y = r\sin t,\ z = t,\ 0 \le t \le a,\ 1 \le r \le 2.$$

Solution. Use the cylindrical coordinates, as in Problem 9.7. The volume multiplier is $(1+r^2)^{1/2}$. The total surface area of this surface is $S = a\int_1^2 (1+r^2)^{1/2}\,dr$, as obtained in Problem 9.7. The coordinates of the centroid satisfy

$$\begin{aligned}
S\,x_0 &= \int x\,d\sigma = \int_0^a \int_1^2 r\cos t\,(1+r^2)^{1/2}\,dr\,dt \\
&= (1/3)(5^{3/2} - 2^{3/2})\sin a, \\
S\,y_0 &= \int y\,d\sigma = \int_0^a \int_1^2 r\sin t\,(1+r^2)^{1/2}\,dr\,dt \\
&= (1/3)(5^{3/2} - 2^{3/2})(1-\cos a), \\
S\,z_0 &= \int z\,d\sigma = \int_0^a \int_1^2 t\,(1+r^2)^{1/2}\,dr\,dt \\
&= (a^2/2)\int_1^2 (1+r^2)^{1/2}\,dr = (a/2)S.
\end{aligned}$$

We see that $z_0 = a/2$. The other coordinates are given as

$$\begin{aligned}
(x_0,\,y_0) &= \frac{1}{3S}(5^{3/2} - 2^{3/2})(\sin a,\,(1-\cos a)) \\
&= \frac{2}{3S}(5^{3/2} - 2^{3/2})\sin\frac{a}{2}(\cos\frac{a}{2},\,\sin\frac{a}{2}).
\end{aligned}$$

9.23 Find the surface area of the torus obtained by rotating the circle

$$(r-2)^2 + z^2 = 1$$

around the z-axis. (See also Problem 8.65.)

Solution. This torus is obtained by rotating a circle of radius one centered at the point $(a,\,b) = (2,\,0)$ in the rz-plane. The arc length of this circle is $\ell = 2\pi$. Hence its surface area is $2\pi a\ell = 2\pi\,2\,2\pi = 8\pi^2$ by Pappus' theorem in Problem 9.22. As an exercise let us compute this area also by a direct integral. A parametric representation for the circle in the rz-plane is $r = 2+\cos\alpha,\ z = \sin\alpha,\ 0 < \alpha < 2\pi$. Hence a parametric representation for the torus is

$$\begin{aligned}
x &= r\cos\theta = (2+\cos\alpha)\cos\theta, \\
y &= r\sin\theta = (2+\cos\alpha)\sin\theta, \\
z &= \sin\alpha,
\end{aligned}$$

where $0 < \alpha < 2\pi$ and $0 < \theta < 2\pi$. The volume multiplier for this representation is obtained in terms of

$$\begin{aligned}
\frac{\partial}{\partial\alpha}(x,\,y,\,z) &= (-\sin\alpha\,\cos\theta,\,-\sin\alpha\,\sin\theta,\,\cos\alpha) = \mathbf{a} \text{ and} \\
\frac{\partial}{\partial\theta}(x,\,y,\,z) &= (-(2+\cos\alpha)\sin\theta,\,(2+\cos\alpha)\cos\theta,\,0) = \mathbf{b}.
\end{aligned}$$

Hence the volume multiplier is the square root of

$$\begin{vmatrix} \langle \mathbf{a}, \mathbf{a} \rangle & \langle \mathbf{a}, \mathbf{b} \rangle \\ \langle \mathbf{b}, \mathbf{a} \rangle & \langle \mathbf{b}, \mathbf{b} \rangle \end{vmatrix} = \begin{vmatrix} 1 & 0 \\ 0 & (2+\cos\alpha)^2 \end{vmatrix} = (2+\cos\alpha)^2.$$

Therefore, the area of the torus is

$$\int_0^{2\pi} \int_0^{2\pi} (2+\cos\alpha) d\alpha\, d\theta = 8\pi^2.$$

9.3 ORIENTED MANIFOLDS

9.25 Let W and Z be oriented spaces. Show that an isomorphism $T : W \to Z$ is orientation-preserving (Definition 9.3.3) if and only if it maps any positive bases of W to a positive basis of Z.

Solution. Let W and Z be oriented by the Euclidean determinants ϱ and ϑ respectively. Let $\dim W = \dim Z = n$ and let $T : W \to Z$ be an isomorphism. Define $\psi : W^n \to \mathbb{R}$ as $\psi(\mathbb{W}) = \vartheta(T\mathbb{W})$, where $\mathbb{W} = (\mathbf{w}_1, \ldots, \mathbf{w}_n) \in W^n$ and $T\mathbb{W} = (T\mathbf{w}_1, \ldots, T\mathbf{w}_n)$. We see that $\psi : W^n \to \mathbb{R}$ is an alternating multilinear function. Then ψ is a multiple of a determinant. Hence there is a $k \in \mathbb{R}$ such that $\psi(\mathbb{W}) = k\,\varrho(\mathbb{W})$ for all $\mathbb{W} \in W^n$. This implies

$$\vartheta(T\mathbb{W}) = k\varrho(\mathbb{W}) \text{ for all } \mathbb{W} \in W^n. \tag{9.2}$$

Assume that $T : W \to Z$ is an orientation-preserving isomorphism. Hence, by Definition 9.3.3, there is a positive basis $\mathbb{E}$ for W such that $T\mathbb{E}$ is a positive basis for Z. This means that $\varrho(\mathbb{E}) > 0$ and $\vartheta(T\mathbb{E}) > 0$. Then $k > 0$. In this case we see that $\vartheta(T\mathbb{U}) > 0$ whenever $\varrho(\mathbb{U}) > 0$. Hence T maps positive bases of W to positive bases of Z. The converse is trivial.

9.27 Show that an isomorphism is orientation-preserving if and only if its inverse is orientation-preserving.

Solution. Let $S : Z \to W$ be the inverse of an isomorphism $T : W \to Z$. With the notations of the solution to Problem 9.25, the relation $\vartheta(T\mathbb{W}) = k\varrho(\mathbb{W})$ for all $\mathbb{W} \in W^n$ is equivalent to $\vartheta(\mathbb{Z}) = k\varrho(S\mathbb{Z})$ for all $\mathbb{Z} \in Z^n$. Then the solution follows.

9.29 Let $T : W \to Z$ be an orientation-preserving isomorphism. Is $-T$ also orientation-preserving? Why?

Solution. If $T : W \to Z$ is an orientation preserving isomorphism, then there is a $k > 0$ that satisfies (9.2). In this case

$$\vartheta((-T)\mathbb{W}) = (-1)^n \vartheta(T\mathbb{W}) = k\varrho(\mathbb{W}) \text{ for all } \mathbb{W} \in W^n.$$

Hence we see that $-T$ is orientation-preserving if and only if $n = \dim Z = \dim W$ is an even integer.

9.31 Let G be a connected set and let $\Psi_i : G \to H_i$ be two charts for a manifold M. Let $T_{\mathbf{m}}$ be the tangent space of M at $\mathbf{m} \in M$. Show that the local orientations $(\Psi_i, \mathbb{E}_i)$ induce either the same orientation on $T_{\mathbf{m}}$ for all $\mathbf{m} \in G \cap M$, or the opposite orientations on $T_{\mathbf{m}}$ for all $\mathbf{m} \in G \cap M$.

Solution. This problem is not correct. A correct version is given below. First we give a counterexample to the statement made in in the version above.

Let Z be the xyz-space and let W be the uvw-space. Let $M = I \cup J$ where

$$\begin{aligned} I &= \{ (x,\, y,\, z) \mid 0 < x < 1,\ y = z = 0 \} \text{ and} \\ J &= \{ (x,\, y,\, z) \mid 0 < x < 1,\ y = 0,\ z = 1 \}. \end{aligned}$$

Let $G = \{ (x,\, y,\, z) \mid 0 < x < 1,\ -\alpha < y < \alpha,\ -\alpha < z < 1 + \alpha \}$ where α is a fixed number, $0 < \alpha < 1$. An easy check shows that M is a manifold and G is an open and connected set containing M. We give two charts $\Psi_i : G \to W$ for M.

These charts take $(x,\, y,\, z) \in G$ to $(u_i,\, v_i,\, w_i)(x,\, y,\, z)$ given as

$$\begin{aligned} u_1(x,\, y,\, z) &= x + 3z \\ v_1(x,\, y,\, z) &= 1 - (1 - y)\cos 2\pi z \\ w_1(x,\, y,\, z) &= (1 - y)\sin 2\pi z \text{ and} \\ u_2(x,\, y,\, z) &= 2 - (2 - x)\cos \pi z \\ v_2(x,\, y,\, z) &= y \\ w_2(x,\, y,\, z) &= (2 - x)\sin \pi z. \end{aligned}$$

We see that $\Psi_i : G \to W$ are $\mathcal{C}^1$ functions. Also, we verify easily that they are one-to-one. The corresponding Jacobian matrices are

$$\begin{aligned} \frac{\partial(u_1,\, v_1,\, w_1)}{\partial(x,\, y,\, z)} &= \begin{bmatrix} 1 & 0 & 3 \\ 0 & \cos 2\pi z & (1-y)\sin 2\pi z \\ 0 & -\sin 2\pi z & (1-y)\cos 2\pi z \end{bmatrix} \text{ and} \\ \frac{\partial(u_2,\, v_2,\, w_2)}{\partial(x,\, y,\, z)} &= \begin{bmatrix} \cos \pi z & 0 & (2-x)\sin \pi z \\ 0 & 1 & 0 \\ -\sin \pi z & 0 & (2-x)\cos \pi z \end{bmatrix}. \end{aligned}$$

We see that their determinants are, respectively, $(1 - y)$ and $(2 - x)$. These are nonzero on G. Hence the derivatives are invertible. This shows that $\Psi_i : G \to W$ are diffeomorphisms. Also, both Ψ_i map I and J, respectively, to

$$\begin{aligned} I' &= \{ (u,\, v,\, w) \mid 0 < u < 1,\ v = w = 0 \} \text{ and} \\ J' &= \{ (u,\, v,\, w) \mid 3 < u < 4,\ v = w = 0 \}. \end{aligned}$$

We see they induce the same orientation on I but reverse orientations on J. Hence the assertion in Problem 9.31 is false. The correct, intended version of this problem is the following.

Problem. Let $\Psi_i : G \to H_i$ be two charts for a manifold M. Let $T_{\mathbf{m}}$ be the tangent space of M at $\mathbf{m} \in M$. If $G \cap M$ is a connected set, then show that the local orientations $(\Psi_i, \mathbb{E}_i)$ induce either the same orientation on $T_{\mathbf{m}}$ for all $\mathbf{m} \in G \cap M$, or the opposite orientations on $T_{\mathbf{m}}$ for all $\mathbf{m} \in G \cap M$.

Solution. We are given that $G \cap M$ is a connected set and $\Psi_i : G \to H_i$ are continuous functions. Then Theorem 4.5.35 shows that $C_i = \Psi_i(G \cap M)$ are also connected sets, as the images of a connected set under continuous functions. Also, since Ψ_is are charts, $C_i = H_i \cap U_i$s are open subsets of the subspaces U_i. Therefore, as stated in Problem 4.83, C_is are arcwise connected.

Now let U_i be oriented by the Euclidean determinant ϖ_i and let $\mathbb{E}_i$ be a positive basis for U_i. Let $\mathbf{m} \in G \cap M$ and $\mathbf{u}_i = \Psi_i(\mathbf{m})$. The orientation induced by Ψ_i on $T_{\mathbf{m}}$ is specified by making the basis $\Phi_i'(\mathbf{u}_i)\mathbb{E}_i$ a positive basis for $T_{\mathbf{m}}$. Here $\Phi_i = \Psi_i^{-1} : H_i \to G$ is the inverse of Ψ_i. We see that the orientations from these two charts are the same if $\Psi_2'(\mathbf{m})(\Phi_1'(\mathbf{u}_1)\mathbb{E}_1)$ is a positive basis for U_2. This is equivalent to the condition that $\varpi_2(\Psi_2'(\mathbf{m})(\Phi_1'(\mathbf{u}_1)\mathbb{E}_1)) > 0$. Here $\mathbf{m} = \Phi_1(\mathbf{u}_1)$.

Define $F : C_1 \to C_2$ by $F(\mathbf{u}_1) = \varpi_2(\Psi_2'(\Phi_1(\mathbf{u}_1))(\Phi_1'(\mathbf{u}_1)\mathbb{E}_1))$, $\mathbf{u}_1 \in U_1$. We see that this is a continuous function. Also, it is nonzero at every $\mathbf{u}_1 \in C_1$, since all transformations involved are isomorphisms and transform linearly independent sets to linearly independent sets. Since C_1 is an arcwise connected set F keeps a fixed sign on C_1. Hence the orientations induced on $T_{\mathbf{m}}$ by these charts are either the same or the opposite at every $\mathbf{m} \in G \cap M$.

9.4 INTEGRALS OF VECTOR FIELDS

9.33 Define $\mathbf{f} : \mathbb{R}^2 \to \mathbb{R}^2$ by $\mathbf{f}(x, y) = (xy, xy^2)$. Compute $\int_C \mathbf{f}$ from the point $(0, 0)$ to the point $(1, 1)$, where C is

1. the parabola $y = x^2$;
2. the circle $x^2 + (y - 1)^2 = 1$, oriented counterclockwise;
3. the circle $x^2 + (y - 1)^2 = 1$, oriented clockwise;
4. the line $y = x$.

Solution. We will use the notations in Definition 9.4.5. The usual unit vectors of the coordinate axes are $\mathbf{i}$ and $\mathbf{j}$. Hence $F(x, y) = xy\mathbf{i} + xy^2\mathbf{j}$.

1. In this case $\mathbf{r}(t) = t\mathbf{i} + t^2\mathbf{j}$, $0 \leq t \leq 1$. Hence

$$\begin{aligned}\int_C \mathbf{f} &= \int_0^1 (t^3\mathbf{i} + t^5\mathbf{j}) \cdot (\mathbf{i} + 2t\mathbf{j})dt \\ &= \int_0^1 (t^3 + 2t^6)dt = (1/4 + 2/7) = 15/28.\end{aligned}$$

2. We have $x(t) = \cos t$, $y = 1 + \sin t$, $-\pi/2 \leq t \leq 0$. Hence

$$\begin{aligned}\int_C \mathbf{f} &= \int_{-\pi/2}^0 (\cos t\,(1 + \sin t)\mathbf{i} + \cos t\,(1 + \sin t)^2\mathbf{j}) \cdot (-\sin t\,\mathbf{i} + \cos t\,\mathbf{j})dt \\ &= \int_{-\pi/2}^0 (-\sin t\,\cos t\,(1 + \sin t) + \cos^2 t\,(1 + \sin t)^2)dt \\ &= 5\pi/16 - 1/2.\end{aligned}$$

3. Take $x(t) = -sint$, $y = 1 - \cos t$, $0 \leq t \leq 3\pi/2$. Then

$$\begin{aligned}\int_C \mathbf{f} &= \int_0^{3\pi/2} (-\sin t\,(1 - \cos t)\mathbf{i} - \sin t\,(1 - \cos t)^2\mathbf{j}) \cdot (-\cos t\,\mathbf{i} + \sin t\,\mathbf{j})dt \\ &= \int_0^{3\pi/2} (\sin t\,\cos t\,(1 + \sin t) - \sin^2 t\,(1 - \cos t)^2)dt \\ &= -15\pi/16 - 1/2.\end{aligned}$$

4. Here $x = y = t$ and $0 \leq t \leq 1$. Hence

$$\begin{aligned}\int_C \mathbf{f} &= \int_0^1 (t^2\mathbf{i} + t^3\mathbf{j}) \cdot (\mathbf{i} + \mathbf{j})dt \\ &= \int_0^1 (t^2 + t^3)dt = (1/3 + 2/4) = 7/12.\end{aligned}$$

9.35 Let A be a connected open set in a Euclidean space Z. Let $\mathbf{f} : A \to Z$ be a continuous vector field. Show that the following are equivalent.

1. There is a $\mathcal{C}^1$ function $F : A \to \mathbb{R}$ such that $\mathbf{f} = \nabla F$.
2. If C is a curve in A, then the line integral $\int_C \mathbf{f}$ depends only on the initial and the final points of C. More explicitly, if C_i are two curves in A with the same initial point P and the same final point Q, then $\int_{C_1} \mathbf{f} = \int_{C_2} \mathbf{f}$.

Solution. Assume that there is a $\mathcal{C}^1$ function $F : A \to Z$ such that $\nabla F = \mathbf{f}$. Let $\overline{I} = [a, b] \subset \mathbb{R}$ be an interval. Let $\mathbf{r} : I \to A$ be a continuous function which is a

$\mathcal{C}^1$ function in the open interval $I = (a, b)$. We claim that

$$\int_a^b \mathbf{f}(\mathbf{r}(t)) \cdot \mathbf{r}'(t)\, dt = F(\mathbf{r}(b)) - F(\mathbf{r}(a)). \tag{9.3}$$

To see this define $G(t) = F(\mathbf{r}(t))$, $t \in \overline{I}$. We see that G is differentiable on I and

$$G'(t) = \nabla F(\mathbf{r}(t)) \cdot \mathbf{r}'(t) = \mathbf{f}(\mathbf{r}(t)) \cdot \mathbf{r}'(t)$$

for all $t \in I$. Then (9.3) follows from the fundamental theorem of calculus, Theorem 8.2.30 and Corollary 8.2.31. This shows that the integral $\int_C \mathbf{f}$ depends only on the initial and the final points of C.

Conversely, assume that if C is a curve in A then $\int_C \mathbf{f}$ depends only on the initial and the final points of C. Choose a point $\mathbf{a}_0 \in A$. Given any other point $\mathbf{a} \in A$ there are finitely many points $\mathbf{a}_i$, $i = 1, \ldots, n$, such that $\mathbf{a}_n = \mathbf{a}$ and that the segment C_i joining $\mathbf{a}_{i-1}$ to $\mathbf{a}_i$ is contained in in A for all i.

9.37 Define $\mathbf{f} : R^3 \to \mathbb{R}^3$ as $\mathbf{f}(x, y, z) = (xyz, xy^2z, z)$. Compute $\int_S \mathbf{f}$ over the following surfaces S, oriented by taking normals with positive z-coordinates.

1. Upper half of the sphere $x^2 + y^2 + z^2 = 1$.
2. The part of the paraboloid $z = 1 - (x^2 + y^2)$.
3. The part of the cone $z = (x^2 + y^2)^{1/2}$ between the planes $z = 1$ and $z = 2$.
4. The helicoidal surface $x = r \cos \vartheta$, $y = r \sin \vartheta$, $z = \vartheta$, where $1 \le r \le 2$ and $0 \le \vartheta \le 2\pi$.

Solution. Let $\mathbf{r}(u, v)$ be a parametric representation for S. The orientation of S induced by this representation is determined by defining $\mathbb{B} = (\mathbf{r}_u(u, v), \mathbf{r}_v(u, v))$ as a positive basis for the tangent space of S at $\mathbf{r}(u, v)$. This orientation agrees with the orientation given in the problem if

$$\mathbf{n}_{\mathbf{r}(u, v)} \cdot \mathbf{r}_u(u, v) \times \mathbf{r}_v(u, v) > 0. \tag{9.4}$$

Here $\mathbf{n}_{\mathbf{r}(u, v)}$ is the unit normal that has a positive z-coordinate. But

$$\mathbf{r}_u(u, v) \times \mathbf{r}_v(u, v) \tag{9.5}$$

is a normal vector of S. Hence (9.4) is satisfied if this vector has a positive z-coordinate. In the first three cases we see that there is a natural parametric representation of the form

$$\mathbf{r}(u, v) = u\mathbf{i} + v\mathbf{j} + w(u, v)\mathbf{k}, \quad (u, v) \in C. \tag{9.6}$$

Here C is the domain of definition of this representation in the uv-plane, **i**, **j**, **k** are the usual unit vectors of the coordinate axes in the xyz-space, and $w : C \to \mathbb{R}$ is a function. In this case an easy computation shows that the z-coordinate of the vector in (9.5) is 1. Therefore the parametric representation in (9.6) induces the required orientation on S. Hence, as in Remarks 9.4.9,

$$\int_S \mathbf{f} = \int_C \mathbf{f}(\mathbf{r}(u,\, v)) \cdot \mathbf{r}_u(u,\, v) \times \mathbf{r}_v(u,\, v)\, du\, dv. \tag{9.7}$$

Let $F(u,\, v) = \mathbf{f}(\mathbf{r}(u,\, v)) \cdot \mathbf{r}_u(u,\, v) \times \mathbf{r}_v(u,\, v)$. The first three cases of this problem amounts to the computation of F and its integral on C.

1. Here $w(u,\, v) = (1 - u^2 - v^2)^{1/2}$ and C is the set $u^2 + v^2 < 1$. Then

$$F(u,\, v) = \begin{vmatrix} uvw & uv^2w & w \\ 1 & 0 & -u/w \\ 0 & 1 & -v/w \end{vmatrix} = u^2v + uv^3 + w.$$

To compute $\int_C F$ introduce polar coordinates $u = r\cos\vartheta$, $v = r\sin\vartheta$. Hence

$$\begin{aligned} \int_C F &= \int_0^1 \int_0^{2\pi} (r^3\cos^2\vartheta\sin\vartheta + r^4\cos\vartheta\sin^3\vartheta + (1 - r^2)^{1/2})\, d\vartheta\, r\, dr \\ &= 2\pi \int_0^1 (1 - r^2)^{1/2} r\, dr = 4\pi/3. \end{aligned}$$

2. This part of the problem is incomplete, as the required part of the paraboloid is not specified. Let us take this part as the part above the xy-plane. Hence $w = 1-(u^2+v^2)$ and C is the set $u^2 + v^2 < 1$. We obtain, as before,

$$\begin{aligned} F(u,\, v) &= \begin{vmatrix} uvw & uv^2w & w \\ 1 & 0 & -2u \\ 0 & 1 & -2v \end{vmatrix} = 2u^2vw + 2uv^3w + w, \text{ and} \\ \int_C F &= \int_0^1 \int_0^{2\pi} F(r\cos\vartheta,\, r\sin\vartheta)\, d\vartheta\, r\, dr \\ &= 2\pi \int_0^1 (1 - r^2) r\, dr = \pi/2. \end{aligned}$$

3. In this case $w = (u^2 + v^2)^{1/2}$ and C is the set $1 < u^2 + v^2 < 4$. Hence

$$\begin{aligned} F(u,\, v) &= \begin{vmatrix} uvw & uv^2w & w \\ 1 & 0 & u/w \\ 0 & 1 & v/w \end{vmatrix} = -u^2v - uv^3 + w, \text{ and} \\ \int_C F &= \int_1^2 \int_0^{2\pi} F(r\cos\vartheta,\, r\sin\vartheta)\, d\vartheta\, r\, dr \\ &= 2\pi \int_1^2 r\, r\, dr = 14\pi/3. \end{aligned}$$

CHAPTER 10

STOKES' THEOREM

10.1 BASIC STOKES' THEOREM

10.1 On the basis of Corollary 10.1.5 one uses $\nabla \cdot \mathbf{f}$ as another notation for

$$\operatorname{div} \mathbf{f} = \operatorname{div} \sum\nolimits_i f_i \mathbf{e}_i = \sum\nolimits_i \frac{\partial f_i}{\partial z_i}.$$

Here $(\mathbf{e}_1, \ldots, \mathbf{e}_n)$ is an orthonormal basis for Z and z_is are the corresponding coordinate functions. If $\mathbf{f} = \nabla F$, then show that $\operatorname{div} \mathbf{f} = \sum_i (\partial^2 F / \partial z_i^2)$. This is also expressed as

$$\operatorname{div} \mathbf{f} = \nabla \cdot \nabla F = \nabla^2 F = \Delta F.$$

The expression $\nabla^2 F = \Delta F$ is called the *Laplacian* of F. A function F is called a *harmonic function* if $\Delta F = 0$.

Solution. If $\mathbf{f} = \nabla F$ then $f_i = (\partial f / \partial x_i)$ and $(\partial f_i / \partial x_i) = (\partial^2 F / \partial x_i^2)$.

Analysis in Vector Spaces.
By M. A. Akcoglu, P. F. A. Bartha and D. M. Ha

10.3 Let Z be an n-dimensional Euclidean space. Let $r = \|\mathbf{z}\|$. Show that $F(\mathbf{z}) = r^{2-n}$ is a harmonic function (Problem 10.1) in $Z \setminus \{\mathbf{0}\}$. Also, if $n = 2$, then show that $F(\mathbf{z}) = \log r$ is a harmonic function in $Z \setminus \{\mathbf{0}\}$.

Solution. With the notations of Problem 10.1, $\partial r/\partial z_i = z_i/r$. If $n = 1$ then $F(\mathbf{z}) = r$ and $dF/dz = z/r$. Hence $(dF/dz)(z) = +1$ for $z > 0$ and $(dF/dz)(z) = -1$ for $z < 0$. Therefore $d^2F/dx^2 = 0$ in both cases. If $n = 2$ then $F(\mathbf{z}) = 1$ and $\nabla^2 F = 0$. If $n \geq 3$, then

$$\begin{aligned}
\frac{\partial F}{\partial z_i} &= (2-n)r^{1-n}\frac{z_i}{r} = (2-n)r^{-n}z_i \text{ and} \\
\frac{\partial^2 F}{\partial z_i^2} &= (2-n)r^{-n} - n(2-n)r^{-n-1}z_i^2/r \\
&= (2-n)r^{-n} - n(2-n)r^{-n-2}z_i^2. \text{ Therefore,} \\
\sum_{i=1}^{n} \frac{\partial^2 F}{\partial z_i^2} &= n(2-n)r^{-n} - n(2-n)r^{-n-2}\sum_{i=1}^{n} z_i^2 \\
&= n(2-n)r^{-n} - n(2-n)r^{-n} = 0.
\end{aligned}$$

If $n = 2$ and $F(\mathbf{z}) = \log r$, then $\partial F/\partial z_i = (1/r)(z_i/r)$ and

$$\frac{\partial^2 F}{\partial z_i^2} = \frac{1}{r^2} - 2\frac{z_i^2}{r^4}.$$

Then $(\partial^2 F/\partial z_1^2) + (\partial^2 F/\partial z_2^2) = 2(1/r^2) - 2(z_1^2 + z_2^2)/r^4 = 0$.

10.2 FLOWS

10.5 Let Z be a Euclidean space. Let J be an open interval containing 0. A function $\Omega : Z \times J \to Z$ is called a *displacement function* if it has the following properties. (1) There is a compact set $K \subset Z$ such that $\Omega(\mathbf{z}, t) = \mathbf{0}$ for $t \in J$ and $\mathbf{z} \notin K$. (2) $\Omega(\mathbf{z}, 0) = \mathbf{0}$ for all $\mathbf{z} \in Z$. (3) $\Omega : (Z \times J) \to Z$ is a $\mathcal{C}^1$ function. (4) The second order mixed partial derivatives

$$\partial(D_Z\Omega)/\partial t \text{ and } D_Z(\partial\Omega/\partial t) \tag{10.1}$$

exist, are equal to each other, and are continuous functions $Z \times J \to L(Z, Z)$. Show that if $F : Z \times J \to Z$ is a smooth flow, then $\Omega(\mathbf{z}, t) = F(\mathbf{z}, t) - \mathbf{z}$ is a displacement function. Conversely, show that if $\Omega(\mathbf{z}, t)$ is a displacement function, then there is an $r > 0$ such that $(-r, r) \subset J$ and such that $F(\mathbf{z}, t) = \mathbf{z} + \Omega(\mathbf{z}, t)$ defines a smooth flow $Z \times (-r, r) \to Z$.

Solution. Let $F : Z \times J \to Z$ be a smooth flow. Hence the conditions in Definitions 10.2.1 and 10.2.4 are satisfied. Let $\Omega(\mathbf{z}, t) = F(\mathbf{z}, t) - \mathbf{z}$. We see that the conditions

(1), (2), and (3) above for Ω follow directly from the corresponding conditions of Definition 10.2.1. Now note that $D_Z\Omega = D_Z F - I$, where $I : Z \to Z$ is the identity. Also, $(\partial\Omega/\partial t) = (\partial F/\partial t)$. Hence

$$\frac{\partial D_Z \Omega}{\partial t} = \frac{\partial D_Z F}{\partial t} \quad \text{and} \quad D_Z \frac{\partial \Omega}{\partial t} = D_Z \frac{\partial F}{\partial t}.$$

Then we see that condition (4) in the definition of displacement functions follows from Definition 10.2.4.

Conversely, let $\Omega : Z \times J \to Z$ be a displacement function. Define

$$F(\mathbf{z}, t) = \mathbf{z} + \Omega(\mathbf{z}, t), \quad (\mathbf{z}, t) \in Z \times J.$$

An easy check shows that F satisfies the first three conditions of Definition 10.2.1 and the smoothness condition in Definition 10.2.4. The only part that remains to prove is Condition (4) in Definition 10.2.1. Hence we will show that there is a $\delta > 0$ such that $(-\delta, \delta) = L \subset J$ and such that $F(\,\cdot\,, t) : Z \to Z$ is a diffeomorphism of Z onto Z for $t \in L$. Let $\xi > 0$ and $[-\xi, \xi] \subset J$. Then

$$D_Z(\partial F/\partial t) = D_Z(\partial\Omega/\partial t) : Z \times [-\xi, \xi] \to L(Z, Z) \tag{10.2}$$

is a bounded function, since it is a continuous function of compact support. Hence there is a $K > 0$ such that

$$\|(\partial(D_Z F)/\partial t)(\mathbf{z}, t)\| = \|D_Z(\partial F/\partial t)(\mathbf{z}, t)\| \le K \tag{10.3}$$

for all $\mathbf{z} \in Z$ and $|t| \le \xi$. Let $0 < \delta < \min(\xi, 1/(2K))$ and $L = (-\delta, \delta)$. Let $t \in L$. We will show that

(1) $D_Z F(\mathbf{z}, t) \in L(Z, Z)$ is invertible,

(2) $F(\,\cdot\,, t) : Z \to Z$ is one-to-one, and

(3) $F(\,\cdot\,, t) : Z \to Z$ maps Z onto Z.

(1) Note that $D_Z F(\mathbf{z}, 0)$ is the identity $I_Z : Z \to Z$, since $F(\mathbf{z}, 0) = \mathbf{z}$ for all $\mathbf{z} \in Z$. Then, by the mean value theorem, Theorem 5.1.13,

$$\begin{aligned} \|D_Z F(\mathbf{z}, t) - I_Z\| &= \|D_Z F(\mathbf{z}, t) - D_Z F(\mathbf{z}, 0)\| && (10.4)\\ &\le K\,|t| < K\,\xi = 1/2. && (10.5) \end{aligned}$$

Hence $D_Z F(\mathbf{z}, t) : Z \to Z$ is invertible by Theorem 4.4.38.

(2) Let $\mathbf{u}, \mathbf{v} \in Z$ and $\mathbf{u} \ne \mathbf{v}$. Then

$$\|(\partial F/\partial t)(\mathbf{v}, t) - (\partial F/\partial t)(\mathbf{u}, t)\| \le K\|\mathbf{v} - \mathbf{u}\| \tag{10.6}$$

by (10.3) and by the mean value theorem. Let $\mathbf{p}(t) = F(\mathbf{v}, t) - F(\mathbf{u}, t)$. Hence $\mathbf{p}(0) = \mathbf{v} - \mathbf{u}$. Also, (10.6) means that $\|\mathbf{p}'(t)\| \le K\|\mathbf{v} - \mathbf{u}\|$ for all $t \in L$ and

$$\begin{aligned} \|\mathbf{p}(t) - \mathbf{p}(0)\| &= \|F(\mathbf{v}, t) - F(\mathbf{u}, t) - \mathbf{v} + \mathbf{u}\| && (10.7) \\ &\le K\|\mathbf{v} - \mathbf{u}\|\,|t|. && (10.8) \end{aligned}$$

If $t \in L$ then $K\,|t| < 1/2$. In this case

$$\begin{aligned} \|F(\mathbf{v}, t) - F(\mathbf{u}, t)\| &\ge \|\mathbf{v} - \mathbf{u}\| - \|F(\mathbf{v}, t) - F(\mathbf{u}, t) - \mathbf{v} + \mathbf{u}\| && (10.9) \\ &\ge \|\mathbf{v} - \mathbf{u}\| - (1/2)\|\mathbf{v} - \mathbf{u}\| && (10.10) \\ &= (1/2)\|\mathbf{v} - \mathbf{u}\| \ne 0. && (10.11) \end{aligned}$$

Hence $F(\,\cdot\,, t) : Z \to Z$ is one-to-one for each $t \in L$.

(3) Now $F(\,\cdot\,, t) : Z \to Z$ is a $\mathcal{C}^1$ function with an invertible derivative at each point. Hence the inverse function theorem applies and shows that this function maps open sets to open sets. Therefore $R = F(Z, t)$ is an open set, since Z is open. We claim that R is also a closed set. Let $\mathbf{q}_n$ be a sequence in R converging to a point $\mathbf{q} \in Z$. Since $F(\,\cdot\,, t)$ is one-to-one there is a unique sequence $\mathbf{p}_n \in Z$ such that $F(\mathbf{p}_n, t) = \mathbf{q}_n$. We see that $\mathbf{p}_n$ is a bounded sequence, since $F(\mathbf{z}, t) = \mathbf{z}$ for all $\mathbf{z}$ outside a compact set $C \subset Z$. Hence by the Bolzano-Weirstrass theorem it has a convergent subspace. Without loss of generality assume that $\mathbf{p}_n \to \mathbf{p}$. Then $\mathbf{q}_n = F(\mathbf{p}_n, t) \to F(\mathbf{p}, t)$, by the continuity of $F(\,\cdot\,, t)$. Since $\mathbf{q}_n \to \mathbf{q}$ we see that $\mathbf{q} = F(\mathbf{p}, t) \in R$. Hence R is closed. Hence R is both open and closed. Since R is not empty we then see that $R = Z$.

To complete the proof note that (2) and (3) imply that $F(\,\cdot\,, t) : Z \to Z$ has an inverse $Z \to Z$ and the inverse function theorem, together with (1), implies that the inverse function is also a $\mathcal{C}^1$ function. Hence $F(\,\cdot\,, t) : Z \to Z$ is a diffeomorphism of Z onto Z for all $t \in L$.

Remarks. Displacement functions are easier to construct than smooth flows. Hence Problem 10.5 gives a convenient way to obtain smooth flows.

10.7 Let (X, Y) be a coordinate system (Definition 3.1.42) in Z with the coordinate projections $P : Z \to X$ and $Q : Z \to Y$. If $F : Z \times J \to Z$ is a smooth flow, then show that there is an open interval I, $0 \in I \subset J$, such that

$$G(\mathbf{x}, t) = PF(\mathbf{x}, t) \quad\text{and}\quad H(\mathbf{y}, t) = QF(\mathbf{y}, t)$$

define smooth flows $G : X \times I \to X$ and $H : Y \times I \to Y$. (Note that here we take $Z = X \oplus Y$ rather than $Z = X \times Y$, so that both $\mathbf{x}$ and $\mathbf{y}$ are in Z.)

Solution. Let $\Omega(\mathbf{z}, t) = F(\mathbf{z}, t) - \mathbf{z}$ be the displacement function of the smooth flow $F : Z \times J \to Z$, as defined in Problem 10.5. An easy check shows that

$$\Omega_X(\mathbf{x}, t) = P\Omega(\mathbf{x}, t) \quad\text{and}\quad \Omega_Y(\mathbf{y}, t) = Q\Omega(\mathbf{y}, t)$$

define two displacement functions $\Omega_X : X \times J \to X$ and $\Omega_Y : Y \times J \to Y$. Hence, as stated in Problem 10.5, there is an open interval I_X such that $0 \in I_X \subset J$ and such that

$$G(\mathbf{x}, t) = \mathbf{x} + \Omega_X(\mathbf{x}, t) = PF(\mathbf{x}, t), \quad (\mathbf{x}, t) \in X \times I_X$$

defines a smooth flow. Similarly, there is an open interval I_Y such that $0 \in I_Y \subset J$ and such that

$$H(\mathbf{y}, t) = \mathbf{y} + \Omega_Y(\mathbf{y}, t) = QF(\mathbf{y}, t), \quad (\mathbf{y}, t) \in Y \times I_Y$$

defines a smooth flow. Then we let $I = I_X \cap I_Y$ to obtain the interval required in the problem.

10.9 Let W be a normed space. Let $\Gamma : Z \times J \to W$ be a function such that $H = (\partial\Gamma/\partial t) : Z \times J \to W$ exists and is continuous. Then show that given a compact set $K \subset Z$ and an $\varepsilon > 0$, there is a $\delta > 0$ such that

$$\|(\Gamma(\mathbf{z}, t) - \Gamma(\mathbf{z}, 0)) - tH(\mathbf{w}, 0)\|_W \leq \varepsilon |t|$$

whenever $|t| + \|\mathbf{z} - \mathbf{w}\| < \delta$ and $\mathbf{z}, \mathbf{w} \in K$. In particular, show that

$$\lim_{t\to 0}(1/t)\,(\Gamma(\mathbf{z}, t) - \Gamma(\mathbf{z}, 0)) = H(\mathbf{z}, 0)$$

uniformly in $\mathbf{z}$ on any compact set $K \subset Z$.

Solution. Let $r > 0$ be such that $I = [-r, r] \subset J$. The choice of the norm on $Z \times \mathbb{R}$ is not important. We may let, for example, $\|(\mathbf{z}, t)\|_{Z\times\mathbb{R}} = \|\mathbf{z}\|_Z + |t|$. The continuous function $H = (\partial\Gamma/\partial t) : Z \times J \to W$ is uniformly continuous on the compact set $K \times I \subset Z \times J \subset Z \times \mathbb{R}$. Find a $\delta > 0$ such that

$$\|H(\mathbf{a}, \alpha) - H(\mathbf{b}, \beta)\|_W \leq \varepsilon$$

whenever $(\mathbf{a}, \alpha), (\mathbf{b}, \beta) \in K \times I$ and $\|(\mathbf{b}, \beta) - (\mathbf{a}, \alpha)\|_{Z\times\mathbb{R}} < \delta$. Let

$$f(s) \quad = \quad \Gamma(\mathbf{z}, s) - \Gamma(\mathbf{z}, 0) - sH(\mathbf{w}, 0)$$

for fixed $\mathbf{z}, \mathbf{w} \in K$ and for $s \in I$. We see that $f : J \to W$ is differentiable and

$$\|f'(s)\|_W = \|H(\mathbf{z}, s) - H(\mathbf{w}, 0)\|_W < \varepsilon$$

if $|s| + \|\mathbf{z} - \mathbf{w}\| \leq \delta$. Then, by the mean value theorem, Theorem 5.1.13,

$$\|f(t) - f(0)\|_W = \|\Gamma(\mathbf{z}, t) - \Gamma(\mathbf{z}, 0) - tH(\mathbf{z}, 0)\|_W \leq \varepsilon\,|t|$$

whenever $|t| + \|\mathbf{z} - \mathbf{w}\| < \delta$. This solves the first part of the problem. The second part follows directly from the first part by letting $\mathbf{w} = \mathbf{z} \in K$.

10.11 Let $F : Z \times J \to Z$ be a smooth flow with the initial velocity field $\mathbf{f}$. Then show that $\lim_{t\to 0}(1/t)(\det D_Z F(\mathbf{z}, t) - 1) = \operatorname{Tr} \mathbf{f}'(\mathbf{z})$ uniformly in $\mathbf{z} \in Z$.

Solution. The flow $F : Z \times J \to Z$ has a compact support C. Hence, if $\mathbf{z} \notin C$ then $F(\mathbf{z}, t) = \mathbf{z}$ for all $t \in J$. Therefore $D_Z F(\mathbf{z}, t) = I$ is the identity on Z for all $\mathbf{z} \notin C$ and for all $t \in J$. In this case $\det D_Z F(\mathbf{z}, t) = 1$ and the result is trivial. Therefore it is enough to prove the uniform convergence for $\mathbf{z} \in C$. This will follow from the an application of the result stated in Problem 10.9. Let $W = \mathbb{R}$ and $\Gamma(\mathbf{z}, t) = \det D_Z F(\mathbf{z}, t)$. Then $\Gamma : Z \times J \to \mathbb{R}$ is the composition of $D_Z F : Z \times J \to L(Z, Z)$ with $\det : L(Z, Z) \to \mathbb{R}$. The determinant function is multi-linear and therefore continuously differentiable. Therefore Γ is a $\mathcal{C}^1$ function as the composition of two $\mathcal{C}^1$ functions. Also,

$$(\partial\Gamma/\partial t)(\mathbf{z}, 0) = \operatorname{Tr} \mathbf{f}'(\mathbf{z}).$$

To obtain this last part apply Theorem C.7.3 with $A = J$ and $a = 0$. Also note that $D_Z F(\,\cdot\,, 0) = I_Z$. Then the result follows directly from Problem 10.9.

10.3 FLUX AND CHANGE OF VOLUME IN A FLOW

10.13 The proof of Theorem 10.3.11 above uses the facts that

$$\lim_{t\to 0}(v(E^t \setminus E) - v(E \setminus E^t))$$

depends only on the initial velocity field and that this dependence is linear. Prove

$$\lim_{t\to 0} (v(E^t \setminus E) - v(E \setminus E^t)) = \int_U \mathbf{n} \cdot \mathbf{f}(\mathbf{u})\, du \tag{10.12}$$

directly, without using this information. Here $F(\mathbf{z}, t) = \mathbf{z} + t\mathbf{f}(\mathbf{z})$.

Solution. It is clear from Theorem 10.3.11 that a factor of $(1/t)$ is missing from the left-hand side of (10.12). The correct form of the relation to be obtained is

$$\lim_{t\to 0} (1/t)(v(E^t \setminus E) - v(E \setminus E^t)) = \int_U \mathbf{n} \cdot \mathbf{f}(\mathbf{u})\, du. \tag{10.13}$$

Let $\mathbf{f} : Z \to Z$ be a $\mathcal{C}^1$ vector field of compact support. Use Theorem 10.2.5 to find an $r_0 > 0$ such that $F(\mathbf{z}, t) = \mathbf{z} + t\,\mathbf{f}(\mathbf{z})$ is a smooth flow $F : Z \times J_0 \to Z$, where $J_0 = (-r_0, r_0)$. Let $U^t = F(U, t)$ be the image of U under this flow. We claim that there is an $r > 0$ such that $J = (-r, r) \subset J_0$ and such that if $t \in J$, then U^t is the graph of a function $f^t : U \to \mathbb{R}$ as $U^t = \{\, \mathbf{u} + f^t(\mathbf{u})\, \mathbf{n} \mid \mathbf{u} \in U \,\}$.

Define a vector field $\mathbf{g} : U \to U$ as $\mathbf{g}(\mathbf{u}) = P\mathbf{f}(\mathbf{u})$, where $P : Z \to U$ is the orthogonal projection on U. Hence $P\mathbf{z} = \mathbf{z} - (\mathbf{z} \cdot \mathbf{n})\, \mathbf{n}$ for all $\mathbf{z} \in Z$. Then

$\mathbf{g} : U \to U$ is a $\mathcal{C}^1$ vector field of compact support. Theorem 10.2.5 shows that there is an $r > 0$ such that $G(\mathbf{u}, t) = \mathbf{u} + t\mathbf{g}(\mathbf{u})$ defines a smooth flow $G : U \times J \to U$ with $J = (-r, r) \subset J_0$. In this case $G(\cdot, t) : U \to U$ is a diffeomorphism of U onto U. Let $H(\cdot, t) : U \to U$ be the inverse diffeomorphism, $t \in J$. We then let

$$f^t(\mathbf{u}) = t\,\mathbf{n} \cdot \mathbf{f}(H(\mathbf{u}, t)), \quad \text{where } \mathbf{u} \in U \text{ and } t \in J.$$

We verify that if $t \in J$ then $U^t \subset Z \cong U \times \mathbb{R}$ is the graph of $f^t : U \to \mathbb{R}$. In fact $\mathbf{z} \in U^t$ means that there is a unique $\mathbf{v} \in U$ such that $\mathbf{z} = \mathbf{v} + t\mathbf{f}(\mathbf{v})$. Hence

$$\mathbf{z} = \mathbf{v} + t\,P\mathbf{f}(\mathbf{v}) + t(\mathbf{f}(\mathbf{v}) \cdot \mathbf{n})\mathbf{n} = G(\mathbf{v}, t) + t(\mathbf{f}(\mathbf{v}) \cdot \mathbf{n})\mathbf{n}.$$

Let $\mathbf{u} = G(\mathbf{v}, t)$. Then $\mathbf{v} = H(\mathbf{u}, t)$ and $\mathbf{z} = \mathbf{u} + f^t(\mathbf{u})\,\mathbf{n}$. This shows that U^t is the graph of $f^t : U \to \mathbb{R}$.

Let $A = \{\, \mathbf{z} \mid \mathbf{z} \cdot \mathbf{n} < 0 \,\}$ and $A^t = F(A, t)$, as before, in the proof of Theorem 10.3.11. We see that, as in (10.50) and (10.50) in that proof,

$$A^t = \{\, (\mathbf{u}, y) \mid \mathbf{u} \in U,\ y < f^t(\mathbf{u}) \,\}$$

and, therefore,

$$\begin{aligned} v(A^t \setminus A) - v(A \setminus A^t) &= \int_U f^t(\mathbf{u})\,d\mathbf{u}, \quad \text{and} \\ (1/t)(v(A^t \setminus A) - v(A \setminus A^t)) &= \int_U \mathbf{f}(H(\mathbf{u}, t)) \cdot \mathbf{n}\,d\mathbf{u}. \end{aligned}$$

Let $C \subset Z$ be a block that contains the support of $\mathbf{f} : Z \to Z$. Let

$$K = \sup\nolimits_{\mathbf{z} \in Z} \|\mathbf{f}(\mathbf{z})\|_Z \quad \text{and} \quad L = \sup\nolimits_{\mathbf{z} \in Z} \|\mathbf{f}'(\mathbf{z})\|_{L(Z,\,Z)}.$$

Let $\mathbf{u} \in U$ and $\mathbf{v} = H(\mathbf{u}, t)$. Then

$$\begin{aligned} \|\mathbf{u} - H(\mathbf{u}, t)\| &= \|\mathbf{u} - \mathbf{v}\| = \|F(\mathbf{v}, t) - \mathbf{v}\| \\ &= \|t\mathbf{f}(\mathbf{v})\| \le |t|\,\|\mathbf{f}(\mathbf{v})\| \le K\,|t|. \end{aligned}$$

Therefore $\|\mathbf{f}(H(\mathbf{u}, t)) - \mathbf{f}(\mathbf{u})\| \le L\,\|H(\mathbf{u}, t) - \mathbf{u}\| \le KL\,|t|$. Hence

$$\left\| \int_U \mathbf{f}(H(\mathbf{u}, t)) \cdot \mathbf{n}\,d\mathbf{u} - \int_U \mathbf{f}(\mathbf{u}) \cdot \mathbf{n}\,d\mathbf{u} \right\| \le KL\,|t|\,v^{n-1}(C \cap U).$$

Here $n = \dim Z$ and v^{n-1} is the $(n-1)$-dimensional volume in U. Therefore

$$\lim_{t \to 0} \int_U \mathbf{f}(H(\mathbf{u}, t)) \cdot \mathbf{n}\,d\mathbf{u} = \int_U \mathbf{f}(\mathbf{u}) \cdot \mathbf{n}\,d\mathbf{u}.$$

Then the solution follows from Lemma 10.3.8, as in the proof of Theorem 10.3.11.

10.4 EXTERIOR DERIVATIVES

10.15 Prove Theorem 10.4.9 for the special cases of $k = 1$ and $k = 2$ by writing out the terms in Equations (10.66)-(10.67) explicitly.

Solution. Let $\omega : G \to \Lambda_1(Z) = L(Z, \mathbb{R})$ be a $\mathcal{C}^1$ tensor field of order one. Let $\mathbf{z} \in G$ and $\mathbf{z}_1, \mathbf{z}_2 \in Z$. Let $\mathbf{w} \in H$ and $\mathbf{w}_1, \mathbf{w}_2 \in W$. Let $\xi = \Phi^*(\omega)$. Then

$$\begin{aligned} \xi(\mathbf{w})(\mathbf{w}_2) &= \Phi^*(\omega)(\mathbf{w})(\mathbf{w}_2) = \omega(\Phi(\mathbf{w}))(\Phi'(\mathbf{w})\mathbf{w}_2). \text{ Hence,} \\ \xi'(\mathbf{w})(\mathbf{w}_1;\, \mathbf{w}_2) &= \omega'(\Phi(\mathbf{w}))(\Phi'(\mathbf{w})\mathbf{w}_1; \Phi'(\mathbf{w})\mathbf{w}_2) \\ &\quad +\omega(\Phi(\mathbf{w})(\Phi''(\mathbf{w})(\mathbf{w}_1; \mathbf{w}_2)). \end{aligned}$$

The last term is symmetric in $\mathbf{w}_1$ and $\mathbf{w}_2$ because of the symmetry of second derivatives. Therefore its antisymmetric part vanishes. Then $d\Phi^*(\omega) = \Phi^*(d\omega)$ follows. If ω is a tensor of second order then

$$\begin{aligned} \xi'(\mathbf{w})(\mathbf{w}_1;\, \mathbf{w}_2,\, \mathbf{w}_3) &= \omega'(\Phi(\mathbf{w}))(\Phi'(\mathbf{w})\mathbf{w}_1; \Phi'(\mathbf{w})\mathbf{w}_2,\, \Phi'(\mathbf{w})\mathbf{w}_3) \\ &\quad +\omega(\Phi(\mathbf{w})(\Phi''(\mathbf{w})(\mathbf{w}_1; \mathbf{w}_2),\, \Phi'(\mathbf{w})\mathbf{w}_3) \\ &\quad +\omega(\Phi(\mathbf{w})(\Phi'(\mathbf{w})\mathbf{w}_2, \Phi''(\mathbf{w})(\mathbf{w}_1; \mathbf{w}_3)). \end{aligned}$$

The antisymmetric parts of the last two terms vanish, because of the symmetry between two of its terms. Then the result follows.

10.17 A tensor field $\xi : G \to \Lambda_1(Z)$ is represented by a vector field $\mathbf{f} : G \to Z$ as in Definition 9.5.6. Hence $\xi(\mathbf{a})(\mathbf{z}) = \mathbf{f}(\mathbf{a}) \cdot \mathbf{z}$ for all $\mathbf{a} \in G$ and $\mathbf{z} \in Z$. What is the application of $d\xi(\mathbf{a})$ to a pair of vectors $(\mathbf{z}_1,\, \mathbf{z}_2) \in Z^2$? Let $(\mathbf{e}_1,\, \dots,\, \mathbf{e}_n)$ be a basis for Z. Let $\mathbf{f} = \sum_i P_i \mathbf{e}_i$. Find $d\xi(\mathbf{a})(\mathbf{e}_i,\, \mathbf{e}_j)$ in terms of P_is.

Solution. We have $\xi'(\mathbf{a})(\mathbf{z}_1; \mathbf{z}_2) = (\mathbf{f}'(\mathbf{a})\mathbf{z}_1) \cdot \mathbf{z}_2$. Here $\mathbf{f}'(\mathbf{a}) \in L(Z,\, Z)$ is the derivative of the vector field $\mathbf{f} : G \to Z$ at $\mathbf{a} \in G$ and $\mathbf{f}'(\mathbf{a})\mathbf{z}_1 \in Z$ is the application of this derivative to $\mathbf{z}_1$. Hence

$$d\xi(\mathbf{a})(\mathbf{z}_1,\, \mathbf{z}_2) = (f'(\mathbf{a})\mathbf{z}_1) \cdot \mathbf{z}_2 - (f'(\mathbf{a})\mathbf{z}_2) \cdot \mathbf{z}_1.$$

If $(\mathbf{e}_1,\, \dots,\, \mathbf{e}_n)$ is a basis for Z then $\mathbf{f}'(\mathbf{a})\mathbf{e}_i = (\partial \mathbf{f}/\partial x_i)(\mathbf{a})$, where x_is are the coordinate functions $Z \to \mathbb{R}$ of this basis. If $\mathbf{f} = \sum_j P_j \mathbf{e}_j$ then

$$\mathbf{f}'(\mathbf{a})\mathbf{e}_i = \sum_k \frac{\partial P_k}{\partial x_i}(\mathbf{a})\mathbf{e}_k \text{ and } (\mathbf{f}'(\mathbf{a})\mathbf{e}_i) \cdot \mathbf{e}_j = \sum_k \frac{\partial P_k}{\partial x_i}(\mathbf{a})(\mathbf{e}_k \cdot \mathbf{e}_j).$$

Therefore we obtain

$$d\xi(\mathbf{a})(\mathbf{e}_i,\, \mathbf{e}_j) = \sum_k \left(\frac{\partial P_k}{\partial x_i}(\mathbf{a})(\mathbf{e}_k \cdot \mathbf{e}_j) - \frac{\partial P_k}{\partial x_j}(\mathbf{a})(\mathbf{e}_k \cdot \mathbf{e}_i) \right).$$

In particular if $(\mathbf{e}_1, \ldots, \mathbf{e}_n)$ is an orthonormal basis then

$$d\xi(\mathbf{a})(\mathbf{e}_i, \mathbf{e}_j) = \frac{\partial P_j}{\partial x_i}(\mathbf{a}) - \frac{\partial P_i}{\partial x_j}(\mathbf{a}).$$

10.19 Let $B = \{ (x, y, z) \mid x = y = 0 \}$ be the z-axis and $G = \mathbb{R}^3 \setminus B$. Define $\mathbf{f} : G \to \mathbb{R}^3$ by $\mathbf{f}(x, y, z) = (x^2 z, y^2 z, x^2 + y^2)$. Define $\Omega : G \to G$ by $\Omega(x, y, z) = (x, y, z - (x^2 + y^2)^{1/2} + (x^2 + y^2))$. Let $\xi : G \to \Lambda_1(\mathbb{R}^3)$ and $\eta : G \to \Lambda_2(\mathbb{R}^3)$ be defined by

$$\xi(\mathbf{a})(\mathbf{z}) = \mathbf{f}(\mathbf{a}) \cdot \mathbf{z} \text{ and } \eta(\mathbf{a})(\mathbf{u}, \mathbf{v}) = \mathbf{f}(\mathbf{a}) \cdot \mathbf{u} \times \mathbf{v}.$$

Compute $d\xi$, $d\eta$, $d\Omega^*(\xi)$, $d\Omega^*(\eta)$, $\Omega^*(d\xi)$, and $\Omega^*(d\eta)$ explicitly and verify Theorem 10.4.9 for these cases.

Solution. Let $(\mathbf{i}, \mathbf{j}, \mathbf{k})$ be the standard orthonormal basis for $\mathbb{R}^3$. Let

$$\mathbf{f} = P\mathbf{i} + Q\mathbf{j} + R\mathbf{k} \text{ with } P(\mathbf{r}) = x^2 z,\ Q(\mathbf{r}) = y^2 z, \text{ and } R(\mathbf{r}) = r^2,$$

where $\mathbf{r} = x\mathbf{i} + y\mathbf{j} + z\mathbf{k}$ and $r = (x^2 + y^2)^{1/2}$. Apply Problem 10.17 to obtain

$$\begin{aligned}
d\xi(\mathbf{r})(\mathbf{i}, \mathbf{j}) &= Q_x - P_y = 0 - 0 = 0,\\
d\xi(\mathbf{r})(\mathbf{i}, \mathbf{k}) &= R_x - P_z = 2x - x^2,\\
d\xi(\mathbf{r})(\mathbf{j}, \mathbf{k}) &= R_y - Q_z = 2y - y^2.
\end{aligned}$$

These functions determine the alternating multilinear function $d\xi : \mathbb{R}^3 \times \mathbb{R}^3 \to \mathbb{R}$. To compute the required pullbacks note that $\Omega(\mathbf{r}) = x\mathbf{i} + y\mathbf{j} + (z - r + r^2)\mathbf{k}$ and

$$\begin{aligned}
\Omega'(\mathbf{r})(\mathbf{i}) &= \mathbf{i} + (2x - (x/r))\mathbf{k},\\
\Omega'(\mathbf{r})(\mathbf{j}) &= \mathbf{j} + (2y - (y/r))\mathbf{k},\\
\Omega'(\mathbf{r})(\mathbf{k}) &= \mathbf{k}. \text{ Hence}\\
\Omega^*(d\xi)(\mathbf{r})(\mathbf{i}, \mathbf{j}) &= d\xi(\Omega(\mathbf{r}))(\Omega'(\mathbf{r})\mathbf{i}, \Omega'(\mathbf{r})\mathbf{j})\\
&= d\xi(\Omega(\mathbf{r}))(\mathbf{i} + (2x - (x/r))\mathbf{k}, \mathbf{j} + (2y - (y/r))\mathbf{k})\\
&= (2y - (y/r))d\xi(\Omega(\mathbf{r}))(\mathbf{i}, \mathbf{k}) - (2x - (x/r))d\xi(\Omega(\mathbf{r}))(\mathbf{j}, \mathbf{k})\\
&= (2y - (y/r))(2x - x^2) - (2x - (x/r))(2y - y^2)\\
&= xy(x - y)((1/r) - 2),\\
\Omega^*(d\xi)(\mathbf{r})(\mathbf{i}, \mathbf{k}) &= d\xi(\Omega(\mathbf{r}))(\Omega'(\mathbf{r})\mathbf{i}, \Omega'(\mathbf{r})\mathbf{k})\\
&= d\xi(\Omega(\mathbf{r}))(\mathbf{i} + (2x - (x/r))\mathbf{k}, \mathbf{k})\\
&= 2x - x^2,\\
\Omega^*(d\xi)(\mathbf{r})(\mathbf{j}, \mathbf{k}) &= d\xi(\Omega(\mathbf{r}))(\Omega'(\mathbf{r})\mathbf{j}, \Omega'(\mathbf{r})\mathbf{k})\\
&= d\xi(\Omega(\mathbf{r}))(\mathbf{j} + (2y - (y/r))\mathbf{k}, \mathbf{k})\\
&= 2y - y^2.
\end{aligned}$$

To specify $\Omega^*(\xi)(\mathbf{r})$ we will compute its applications to **i**, **j**, and **k** as

$$(\Omega^*(\xi)(\mathbf{r}))(\mathbf{i}) = \widetilde{P}(\mathbf{r}),\ \ (\Omega^*(\xi)(\mathbf{r}))(\mathbf{j}) = \widetilde{Q}(\mathbf{r}),\ \ (\Omega^*(\xi)(\mathbf{r}))(\mathbf{k}) = \widetilde{R}(\mathbf{r}).$$

We have

$$\begin{aligned}
(\Omega^*(\xi))(\mathbf{r})(\mathbf{i}) &= \xi(\Omega(\mathbf{r}))(\Omega'(\mathbf{i}))\\
&= \xi(\Omega(\mathbf{r}))(\mathbf{i} + (2x - (x/r))\mathbf{k})\\
&= \xi(\Omega(\mathbf{r}))(\mathbf{i}) + (2x - (x/r))\xi(\Omega(\mathbf{r}))(\mathbf{k})\\
&= P(\Omega(\mathbf{r})) + (2x - (x/r))R(\Omega(\mathbf{r}))\\
&= x^2(z - r + r^2) + (2x - (x/r))r^2 = \widetilde{P}(\mathbf{r})\\
(\Omega^*(\xi))(\mathbf{r})(\mathbf{j}) &= y^2(z - r + r^2) + (2y - (y/r))r^2 = \widetilde{Q}(\mathbf{r})\\
(\Omega^*(\xi))(\mathbf{r})(\mathbf{k}) &= r^2 = \widetilde{R}(\mathbf{r}).
\end{aligned}$$

Hence we obtain

$$\begin{aligned}
d(\Omega^*(\xi))(\mathbf{r})(\mathbf{i},\, \mathbf{j}) &= (\widetilde{Q}_x - \widetilde{P}_y)(\mathbf{r}) = xy(x - y)((1/r) - 2),\\
d(\Omega^*(\xi))(\mathbf{r})(\mathbf{i},\, \mathbf{k}) &= (\widetilde{R}_x - \widetilde{P}_z)(\mathbf{r}) = 2x - x^2,\\
d(\Omega^*(\xi))(\mathbf{r})(\mathbf{j},\, \mathbf{k}) &= (\widetilde{R}_y - \widetilde{Q}_z)(\mathbf{r}) = 2y - y^2.
\end{aligned}$$

This shows that $d(\Omega^*(\xi)) = \Omega^*(d\xi)$. Similarly we see that

$$\begin{aligned}
\eta(\mathbf{r})(\mathbf{i},\, \mathbf{j}) &= \mathbf{f}(\mathbf{r}) \cdot \mathbf{k} = r^2,\\
\eta(\mathbf{r})(\mathbf{i},\, \mathbf{k}) &= \mathbf{f}(\mathbf{r}) \cdot (-\mathbf{j}) = -y^2 z,\\
\eta(\mathbf{r})(\mathbf{j},\, \mathbf{k}) &= \mathbf{f}(\mathbf{r}) \cdot (\mathbf{i}) = x^2 z.
\end{aligned}$$

Therefore $\eta'(\mathbf{r})(\mathbf{k};\, \mathbf{i},\, \mathbf{j}) = 0$, $\eta'(\mathbf{r})(\mathbf{i};\, \mathbf{j},\, \mathbf{k}) = 2xz$, $\eta'(\mathbf{r})(\mathbf{j};\, \mathbf{k},\, \mathbf{i}) = 2yz$. Hence $d\eta$ is obtained as $(d\eta)(\mathbf{r})(\mathbf{i},\, \mathbf{j},\, \mathbf{k}) = 2(x + y)z$. Hence

$$\begin{aligned}
(\Omega^* d\eta)(\mathbf{r})(\mathbf{i},\, \mathbf{j},\, \mathbf{k}) &= (d\eta)(\Omega(\mathbf{r}))(\Omega'(\mathbf{r})\mathbf{i},\, \Omega'(\mathbf{r})\mathbf{j},\, \Omega'(\mathbf{r})\mathbf{k})\\
&= (d\eta)(\Omega(\mathbf{r}))(\mathbf{i} + (2x - (x/r))\mathbf{k},\, \mathbf{j} + (2y - (y/r))\mathbf{k},\, \mathbf{k})\\
&= (d\eta)(\Omega(\mathbf{r}))(\mathbf{i},\, \mathbf{j},\, \mathbf{k})\\
&= 2(x + y)(z - r + r^2).
\end{aligned}$$

On the other hand $(\Omega^*\eta)(\mathbf{r})(\mathbf{u},\, \mathbf{v}) = \eta(\Omega(\mathbf{r}))(\Omega'(\mathbf{r})\mathbf{u},\, \Omega'(\mathbf{r})\mathbf{v})$ is given as

$$\begin{aligned}
(\Omega^*\eta)(\mathbf{r})(\mathbf{i},\, \mathbf{j}) &= r^2 + ((1/r) - 2)(x^3 + y^3)(z - r + r^2),\\
(\Omega^*\eta)(\mathbf{r})(\mathbf{k},\, \mathbf{i}) &= y^2(z - r + r^2),\\
(\Omega^*\eta)(\mathbf{r})(\mathbf{j},\, \mathbf{k}) &= x^2(z - r + r^2).
\end{aligned}$$

To obtain $(d(\Omega^*\eta))(\mathbf{r})(\mathbf{i},\, \mathbf{j},\, \mathbf{k})$ we compute the necessary derivatives as

$$\begin{aligned}
(\Omega^*\eta)'(\mathbf{r})(\mathbf{i};\, \mathbf{j},\, \mathbf{k}) &= 2x(z - r + r^2) + x^2(-(x/r) + 2x),\\
(\Omega^*\eta)'(\mathbf{r})(\mathbf{j};\, \mathbf{k},\, \mathbf{i}) &= 2y(z - r + r^2) + x^2(-(y/r) + 2y),\\
(\Omega^*\eta)'(\mathbf{r})(\mathbf{k};\, \mathbf{i},\, \mathbf{j}) &= ((1/r) - 2)(x^3 + y^3).
\end{aligned}$$

Hence $(\Omega^*\eta)'(\mathbf{r})(\mathbf{i}, \mathbf{j}, \mathbf{k}) = 2(x+y)(z-r+r^2)$ and $d(\Omega^*(\eta)) = \Omega^*(d\eta)$ is verified.

10.5 REGULAR AND ALMOST REGULAR SETS

10.21 Let $F : Z \times J \to Z$ be a general smooth flow with an initial velocity field $\mathbf{f}$ with a compact support contained in an open set G. Assume that G is a $\mathcal{C}^1$ regular neighborhood of a set E and repeat Problem 10.20 for this case.

Solution. The proof of Stokes' theorem given in the text is valid only for $\mathcal{C}^2$ boundaries. The solution below outlines a general proof for $\mathcal{C}^1$ boundaries.

1. If G is a $\mathcal{C}^1$ regular neighborhood of E, then $G \cap \partial E$ is a $\mathcal{C}^1$ surface. Given $\mathbf{m} \in \partial E$ there is a neighborhood G_0 such that $\mathbf{m} \in G_0 \subset G$ and such that $M = G_0 \cap \partial E$ is a graph with respect to an orthogonal coordinate system in Z. We specify this coordinate system as $(X, \mathbb{R})$ so that $Z \cong X \times \mathbb{R}$. The points $\mathbf{z} \in Z$ are denoted as $\mathbf{z} = (\mathbf{x}, y) \in X \times \mathbb{R}$. Let $P : Z \to X$ be the orthogonal projection on X and let $\mathbf{e} = (\mathbf{0}, 1) \in Z$ be the unit vector of the $\mathbb{R}$-axis. If $\mathbf{z} = (\mathbf{x}, y)$ then $y = \mathbf{e} \cdot \mathbf{z}$ and $\mathbf{x} = P\mathbf{z} - y\,\mathbf{e}$. Assume that M is a graph in this system. Then there is an open set $C \subset X$ and a $\mathcal{C}^1$ function $h : C \to \mathbb{R}$ such that

$$M = \{\, (\mathbf{x}, y) \mid \mathbf{x} \in C \text{ and } y = h(\mathbf{x}) \,\} \subset X \times \mathbb{R} \cong Z.$$

The set E is on one side of M, since G is a regular neighborhood of E. Without loss of generality assume that $G_0 \cap E$ is in the region below the boundary. Hence

$$G_0 \cap E^o \subset A = \{\, (\mathbf{x}, y) \mid \mathbf{x} \in C \text{ and } y < h(\mathbf{x}) \,\}.$$

2. Let $\mathbf{f} : Z \to Z$ be a continuous vector field with a compact support K contained in G_0. We claim that $\int_{\partial E} \mathbf{f}$ is expressed in terms of $\mathbf{f}$ and h as

$$\int_{\partial E} \mathbf{f} = \int_X \mathbf{f}(\mathbf{x}, h(\mathbf{x})) \cdot (\mathbf{e} - \nabla h(\mathbf{x}))\, d\mathbf{x}. \tag{10.14}$$

In fact, the tangent space of ∂E at $\mathbf{m} = (\mathbf{a}, h(\mathbf{a}))$ is given as

$$y - h(\mathbf{a}) = h'(\mathbf{a})(\mathbf{x} - \mathbf{a}) = \nabla h(\mathbf{a}) \cdot (\mathbf{x} - \mathbf{a}). \tag{10.15}$$

Now $(\mathbf{z} - \mathbf{m}) \cdot \mathbf{e} = y - h(\mathbf{a})$. Also,

$$\nabla h(\mathbf{a}) \cdot (\mathbf{z} - \mathbf{m}) = \nabla h(\mathbf{a}) \cdot (\mathbf{x} - \mathbf{a})$$

since $\nabla h(\mathbf{a}) \in X$. Hence (10.15) can be also written as

$$(\mathbf{e} - \nabla h(\mathbf{a})) \cdot (\mathbf{z} - \mathbf{m}) = 0.$$

Therefore $\mathbf{n} = (\mathbf{e} - \nabla h(\mathbf{a}))$ is a normal vector of ∂E at $\mathbf{m}$. This is an outer normal, since $\mathbf{n} \cdot \mathbf{e} > 0$. Let $\mathbf{n}_0 = \mathbf{n}/\|\mathbf{n}\|$ be the unit normal vector. Now, by Lemma 9.6.6,

$$\int_{\partial E} \mathbf{f} = \int_X \mathbf{f}(\mathbf{x},\, h(\mathbf{x})) \cdot \mathbf{n}_0(\mathbf{x},\, h(\mathbf{x}))/\rho(P_{\mathbf{x}})\, d\mathbf{x}.$$

Here $\rho(P_x)$ is the volume multiplier of the orthogonal projection $P_x : T_{\mathbf{x}} \to X$ of the tangent space $T_{\mathbf{x}}$ of ∂E at $(\mathbf{x},\, h(\mathbf{x}))$ onto X. By the arguments in Remarks 9.6.8 it follows that $\rho(P_{\mathbf{x}}) = \mathbf{e} \cdot \mathbf{n}_0(\mathbf{x},\, h(\mathbf{x}))$. Hence

$$\rho(P_{\mathbf{x}}) = \mathbf{e}(\mathbf{e} - \nabla h(\mathbf{x}))/\|\mathbf{n}(\mathbf{x},\, h(\mathbf{x}))\| = 1/\|\mathbf{n}(\mathbf{x},\, h(\mathbf{x}))\|$$

since $\mathbf{e} \perp \nabla h(\mathbf{x})$. Then we see that (10.14) follows from

$$\begin{aligned} \mathbf{f}(\mathbf{x},\, h(\mathbf{x})) \cdot \mathbf{n}_0(\mathbf{x},\, h(\mathbf{x}))/\rho(P_{\mathbf{x}}) &= \mathbf{f}(\mathbf{x},\, h(\mathbf{x})) \cdot \mathbf{n}(\mathbf{x},\, h(\mathbf{x})) \\ &= \mathbf{f}(\mathbf{x},\, h(\mathbf{x})) \cdot (\mathbf{e} - \nabla h(\mathbf{x})). \end{aligned}$$

3. Let $F : Z \times J \to Z$ be a smooth flow with a compact support K and with the initial velocity field $\mathbf{f} : Z \to Z$. Without loss of generality assume that $K \subset G_0$. The general case is obtained by the partitions of unity, Theorem D.1.8. As in the proof of Theorem 10.3.11, we see that the solution would follow if we can show that

$$\lim_{t \to 0} \frac{1}{t} \left(v(A^t \setminus A) - v(A \setminus A^t) \right) = \int_{\partial E} \mathbf{f}. \tag{10.16}$$

Here $A^t = F(A,\, t)$ is the image of A under the flow F. To obtain (10.16) we proceed as in the solution of Problem 10.13. First we show that, if $|t|$ is sufficiently small, then $M^t = F(M,\, t)$ is also the graph of a $\mathcal{C}^1$ function $h^t : C \to \mathbb{R}$. To see this define $H : C \to C \times \mathbb{R} \subset Z$ as, with $\mathbf{x} \in C$, $\mathbf{a} \in X$,

$$H(\mathbf{x}) = (\mathbf{x},\, h(\mathbf{x})), \quad H'(\mathbf{x})\mathbf{a} = (\mathbf{a},\, \nabla h(\mathbf{x}) \cdot \mathbf{a}),$$

We will show that if $\mathbf{x} \in C$ and if $|t|$ is sufficiently small then there is a unique vector $\mathbf{u} = \mathbf{u}(\mathbf{x},\, t) \in C$, that depends on $\mathbf{x}$ and t, such that $PF(H(\mathbf{u}),\, t) = \mathbf{x}$. In this case we see that $M^t = F(M,\, t)$ is the graph of the function

$$h^t(\mathbf{x}) = (F(H(\mathbf{u}(\mathbf{x},\, t)),\, t) - \mathbf{x}) \cdot \mathbf{e}. \tag{10.17}$$

The existence and the differentiability of $\mathbf{u}$ follows from the implicit function theorem, Theorem 6.4.6. In fact $\mathbf{u}$ is obtained as the solution of the equation

$$\vartheta(\mathbf{u},\, \mathbf{x},\, t) = PF(H(\mathbf{u}),\, t) - \mathbf{x} = \mathbf{0}.$$

Here ϑ is defined on the open set $C \times C \times J \subset X \times X \times \mathbb{R}$ and takes values in X. We see that $\vartheta : C \times C \times J \to X$ is a $\mathcal{C}^1$ function and $\vartheta(\mathbf{x},\, \mathbf{x},\, 0) = \mathbf{0}$. Also, the derivative of ϑ with respect to $\mathbf{u}$ applied to a vector $\mathbf{a} \in X$ is

$$PD_Z F(H(\mathbf{u}),\, t) H'(\mathbf{u})\mathbf{a} = PD_Z F(H(\mathbf{u}),\, t)(\mathbf{a},\, \nabla H(\mathbf{u}) \cdot \mathbf{a}).$$

At $t = 0$ this derivative is the identity $X \to X$. Hence, by continuity, it is an invertible linear transformation for sufficiently small $|t|$. Then the implicit function theorem gives the existence of a unique $\mathcal{C}^1$ function $\mathbf{u}(\mathbf{x}, t)$, defined for $\mathbf{x} \in C$ and for sufficiently small $|t|$ such that $PF(H(\mathbf{u}(\mathbf{x}, t)), t) = \mathbf{x}$. In terms of this function $h^t : C \to \mathbb{R}$ is given by (10.17). Also, $h^0 = h$. Hence

$$\lim_{t\to 0} \frac{1}{t} \left(v(A^t \setminus A) - v(A \setminus A^t)\right) = \lim_{t\to 0} \frac{1}{t} \left(h^t(\mathbf{x}) - h(\mathbf{x})\right) d\mathbf{x}. \tag{10.18}$$

Denote $h^t(\mathbf{x})$ as

$$h^t(\mathbf{x}) = \varphi(\mathbf{x}, t) = (F(H(\mathbf{u}(\mathbf{x}, t)), t) - \mathbf{x}) \cdot \mathbf{e}. \tag{10.19}$$

As this is a $\mathcal{C}^1$ function of $(\mathbf{x}, t)$ the ratio

$$(1/t)(\varphi(\mathbf{x}, t) - \varphi(\mathbf{x}, 0))$$

converges of $(\partial\varphi/\partial t)(\mathbf{x}, 0)$ uniformly on compact sets in C. Therefore we see that the limit in (10.18) is $\int_C (\partial\varphi/\partial t)(\mathbf{x}, 0)\, d\mathbf{x}$.

4. Finally we compute $(\partial\varphi/\partial t)(\mathbf{x}, 0)$ from its expression in (10.19). We have

$$\begin{aligned}
\frac{\partial\varphi}{\partial t}(\mathbf{x}, 0) &= \mathbf{e} \cdot \Big(D_Z F(H(\mathbf{u}(\mathbf{x}, 0)), 0) H'(\mathbf{u}(\mathbf{x}, 0)) \frac{\partial \mathbf{u}}{\partial t}(\mathbf{x}, 0) \\
&\quad + \frac{\partial F}{\partial t}(H(\mathbf{u}(\mathbf{x}, 0)), 0)\Big) \\
&= \mathbf{e} \cdot \Big(H'(\mathbf{x}) \frac{\partial \mathbf{u}}{\partial t}(\mathbf{x}, 0) + \mathbf{f}(H(\mathbf{x}))\Big) \\
&= \nabla h(\mathbf{x}) \cdot \frac{\partial \mathbf{u}}{\partial t}(\mathbf{x}, 0) + \mathbf{e} \cdot \mathbf{f}(H(\mathbf{x})).
\end{aligned}$$

We compute $(\partial\mathbf{u}/\partial t)(\mathbf{x}, 0)$ from the defining equation

$$PF(H(\mathbf{u}(\mathbf{x}, t)), t) = \mathbf{x}.$$

Differentiation with respect to t gives

$$PD_Z F(H(\mathbf{u}(\mathbf{x}, t)), t) H'(\mathbf{u}(\mathbf{x}, t)) \frac{\partial \mathbf{u}}{\partial t}(\mathbf{x}, t) + P\frac{\partial F}{\partial t}(H(\mathbf{u}(\mathbf{x}, t)), t) = \mathbf{0}.$$

By evaluation this equation at $t = 0$ we obtain

$$PH'(\mathbf{x}) \frac{\partial \mathbf{u}}{\partial t}(\mathbf{x}, 0) + P\mathbf{f}(H(\mathbf{x})) = \mathbf{0}.$$

This gives $(\partial\mathbf{u}/\partial t)(\mathbf{x}, 0) = -P\mathbf{f}(H(\mathbf{x}))$, since $(\partial\mathbf{u}/\partial t)(\mathbf{x}, 0) \in X$. Hence

$$\begin{aligned}
\mathbf{e} \cdot H'(\mathbf{x}) \frac{\partial \mathbf{u}}{\partial t}(\mathbf{x}, 0) &= -\mathbf{e}(P(\mathbf{f}(H(\mathbf{x})), \nabla h(\mathbf{x}) \cdot P(\mathbf{f}(H(\mathbf{x}))) \\
&= -\nabla h(\mathbf{x}) \cdot P(\mathbf{f}(H(\mathbf{x}))) = -\nabla h(\mathbf{x}) \cdot \mathbf{f}(H(\mathbf{x})).
\end{aligned}$$

This gives $(\partial\varphi/\partial t)(0, t) = \mathbf{f}(H(\mathbf{x}) \cdot (\mathbf{e} - \nabla h(\mathbf{x})$. This agrees with (10.14). Hence the solution is completed.

10.23 Compute $\int_S \mathbf{f}$ where $\mathbf{f}$ is as in Problem 10.22 and S is the upper half $(z > 0)$ of the unit sphere in $\mathbb{R}^3$ oriented by the outer normals.

Solution. We see that $E = \{ (x, y, z) \mid 0 < z < (1 - x^2 - y^2)^{1/2} \}$ is an almost regular region. (See the solution of Problem 10.25 for more details.) Apply Stokes' theorem to E. We have $\nabla \cdot \mathbf{f}(x, y, z) = 2x + 2y + 1$. Then $\int_E \nabla \mathbf{f} = 2\pi/3$ by an easy computation. Hence $\int_{\partial E} \mathbf{f} = \int_S \mathbf{f} + \int_D \mathbf{f} = 2\pi/3$, where S is as given in the problem and D is the unit disc in the xy-plane oriented by the outer normal of E. We see that this outer normal is $-\mathbf{k}$ for D. Hence

$$\int_D \mathbf{f} = -\int_D \mathbf{f}(x, y, 0) \cdot \mathbf{k}\, dx\, dy = \int_0^{2\pi} \int_0^1 e^{r^2} r\, dr\, d\vartheta = (1 - e)\pi.$$

Therefore $\int_S \mathbf{f} = 2\pi/3 - (1 - e)\pi = (3e - 1)\pi/3$.

10.25 Let $a > 0$. Let E be the region in $\mathbb{R}^3$ specified by $x^2 + y^2 < z < a$. Show that E is an almost regular set. Compute $\int_{\partial E} \mathbf{f}$, where $\mathbf{f}$ is as in Problem 10.22 and the integral on ∂E is as defined in Definition 10.5.24.

Solution. The boundary of E consists of the surfaces $z = x^2 + y^2$, $z < a$, and $x^2 + y^2 < a$, $z = a$ and also of the circle C given as $x^2 + y^2 = a$, $z = a$. Also, E is always on one side of the surface part of the boundary. We see the that the enlargement of C by $s > 0$ is a torus C_s. This torus is obtained by rotating the open disc $(z - a)^2 + (r - \sqrt{a})^2 < s^2$ in the rz-plane about the z-axis. Here r and z are the usual cylindrical coordinates. The volume of C_s is $2\pi^2\sqrt{a}\, s^2$. This follows by an easy direct computation or from Pappus theorem, Problem 8.64. We see that $v(C_s)/s \to 0$ as $s \to 0^+$ and also that $(\partial E) \setminus C_s$ is a compact subset of the regular part of the boundary. Therefore E is an almost regular set.

The integral $\int_{\partial E} \mathbf{f}$ is easily computed as $\int_E \nabla \cdot \mathbf{f}$. We have $\nabla \cdot \mathbf{f}(x, y, z) = 2x + 2y + 1$ from the solution of Problem 10.23. Hence, in cylindrical coordinates,

$$\begin{aligned}
\int_{\partial E} \mathbf{f} &= \int_0^{\sqrt{a}} \int_0^{2\pi} \int_{r^2}^{a} (2r\cos\vartheta + 2r\sin\vartheta + 1) r\, dz\, d\vartheta\, dr \\
&= \int_0^{\sqrt{a}} \int_0^{2\pi} (2r\cos\vartheta + 2r\sin\vartheta + 1)(a - r^2) r\, d\vartheta\, dr \\
&= 2\pi \int_0^{\sqrt{a}} (a - r^2) r\, dr = \pi a^2/2.
\end{aligned}$$

10.27 Let E ba an almost regular set in Z. Let F and G be two real-valued $\mathcal{C}^2$ functions on Z. Show that

$$\int_{\partial E} (F\nabla G - G\nabla F) = \int_E (F\nabla^2 G - G\nabla^2 F).$$

Solution. This follows directly from Problem 10.26, which, in turn, is a consequence of Stokes' theorem and the computation of $\nabla \cdot (F\nabla G)$ by the product rule of differentiation.

10.29 Let E be an almost regular set in $\mathbb{R}^3$ containing the origin in its interior. Compute $\int_{\partial E} \nabla F$ where F is as in Problem 10.28 and the integral on ∂E is as defined in Definition 10.5.24.

Solution. As stated in Problem 10.3, $F = (x^2+y^2+z^2)^{-1/2}$ is a harmonic function in $\mathbb{R}^3 \setminus \{\mathbf{0}\}$. Since $\mathbf{0} \in E^o$ there is an $s > 0$ such that $B_{2s}(\mathbf{0})$, the ball of radius $2s$ about the origin, is contained in E^o. Let $H = E \setminus B_s(\mathbf{0})$. We see that H is an almost regular set and $\nabla \cdot \nabla F(\mathbf{v}) = 0$ for all $\mathbf{v} \in H$. Therefore

$$\int_{\partial H} \nabla F = \int_H \nabla \cdot \nabla F(\mathbf{v})\, d\mathbf{v} = 0.$$

But $\partial H = (\partial E) \cup (\partial B_s(\mathbf{0}))$ ∂H is oriented by the outer normal with respect to H. This coincides with the outer normal of E with respect to E. But the outer normal of $\partial B_s(\mathbf{0})$ with respect to H is the inner normal with respect to $B_s(\mathbf{0})$. Let S be the boundary of $B_s(\mathbf{0})$ oriented by the outer normal of this ball. Then

$$\begin{aligned}\int_{\partial H} \nabla F &= \int_{\partial E} \nabla F - \int_S \nabla F = 0 \text{ and, therefore,}\\ \int_{\partial E} \nabla F &= \int_S \nabla F.\end{aligned}$$

The last integral can be computed easily. We see that, with $r = (x^2+y^2+z^2)^{1/2}$,

$$\nabla F(x,\, y,\, z) = -r^{-3}(x,\, y,\, z) \text{ and } \mathbf{n} = r^{-1}(x,\, y,\, z),$$

where $\mathbf{n}$ is the outer unit normal vector of $B_r(\mathbf{0})$. Hence on the surface of $B_s(\mathbf{0})$,

$$(\nabla F \cdot \mathbf{n}) = -s^{-4}(x^2+y^2+z^2) = -s^{-2}$$

is a constant. Therefore the integral of this function on the surface of $B_s(\mathbf{0})$ is

$$\int_S (\nabla F \cdot \mathbf{n})\, d\sigma = -s^{-2}(4\pi s^2) = -4\pi.$$

Here $4\pi s^2$ is the surface area of the sphere of radius s, as obtained in Example 9.6.19. Therefore $\int_{\partial E} \nabla F = -4\pi$ whenever $\mathbf{0} \in E^o$.

10.31 Let Z be an n-dimensional Euclidean space. Let $F(\mathbf{z}) = \|\mathbf{z}\|^{2-n}$. Compute $\int_S \nabla F$ where S is the unit sphere $\|\mathbf{z}\| = 1$ in Z oriented by the outer normals.

Solution. We proceed as in the last computation in the solution of Problem 10.29. Let $r = (z_1^2 + \cdots + z_n^2)^{1/2} = \|\mathbf{z}\|$. Hence

$$\begin{aligned} \nabla r^{(2-n)} &= (2-n)r^{1-n}\,(1/r)(z_1, \ldots, z_n) \text{ and} \\ \mathbf{n}(\mathbf{z}) &= (1/r)(z_1, \ldots, z_n). \end{aligned}$$

Here $\mathbf{n}(\mathbf{z})$ is the outer unit normal vector of the sphere of radius r at a point $\mathbf{z}$ on this sphere. Hence $\nabla F(\mathbf{z}) \cdot \mathbf{n}(\mathbf{z}) = (2-n)r^{1-n}$ is constant on this sphere and

$$\int_S \nabla F \cdot \mathbf{n}\, d\sigma = (2-n)r^{1-n}(nr^{n-1}S_n) = n(2-n)S_n.$$

Here S_n is the volume of the unit ball in $\mathbb{R}^n$ and $nr^{n-1}S_n$ is the surface area of a ball of radius r in $\mathbb{R}^n$. This was obtained in Example 9.6.19.

10.33 Repeat Problem 10.32 under the assumption that $\mathbf{0} \notin \overline{E}$.

Solution. As stated in Problem 10.3, $F = \|\mathbf{z}\|^{2-n}$ is a harmonic function in $\mathbb{R}^n \setminus \{\mathbf{0}\}$. Since $\mathbf{0} \notin \overline{E}$ we see that $\nabla \cdot \nabla F = 0$ on E. Hence

$$\int_{\partial E} \nabla F = \int_E \nabla \cdot \nabla F = 0.$$

10.35 Let $F(x, y) = \log(x^2 + y^2)$. Let C be the unit circle $x^2 + y^2 = 1$. Note that C is both a surface and a line in $\mathbb{R}^2$. (See also Problem 9.38.) Compute the surface integral $\int_C \nabla F$ where C is oriented by the outer normals.

Solution. We have $\nabla F(x, y) = 2r^{-2}(x, y)$ with $r = (x^2 + y^2)^{1/2}$. The outer unit normal vector of C at $(x, y) \in C$ is $\mathbf{n}(x, y) = (x, y)$. Hence

$$\int_C \nabla F \cdot \mathbf{n} = \int_C 2 = 2(2\pi) = 4\pi.$$

10.37 Repeat Problem 10.36 under the assumption that $\mathbf{0} \notin \overline{E}$.

Solution. We have $\nabla \cdot \nabla F = 0$ on E. In fact, as stated in Problem 10.3, F is harmonic in $\mathbb{R}^2 \setminus \mathbf{0}$ and $\mathbf{0} \notin \overline{E}$. Hence $\int_{\partial E} \nabla F = \int_E \nabla \cdot \nabla F = 0$.

10.39 Repeat Problem 10.38 under the assumption that $\mathbf{0} \notin \overline{E}$.

Solution. Let $\mathbf{r}(t) = (x(t), y(t))$ be a parametric representation of C. If $\mathbf{f} = (P, Q)$ then the line integral of $\mathbf{f}$ on C is $\int_C (Pdx, Qdy)$ and the surface integral of $\mathbf{f}$ on C is

$\int_C(-Qdy+Pdx)$. Hence the required integral is the surface integral of $\mathbf{g} = (y,\, x)/r$. We have div $\mathbf{g} = -xy/r^3$. This a continuous function on E and Stokes' theorem is applicable. Hence we see that $\int_C (1/r)(-xdx + ydy) = -\int_E xy/r^3\, dxdy$.

10.41 Let F and E be as in Problem 10.40. Consider the boundary-surface $\mathcal{S}_E$ as a curve C, oriented by the outer normals according to the right-hand rule. Consider $\mathbb{R}^2$ as the xy-plane in the xyz-space. Let $\mathbf{k} = (0,\, 0\, 1)$ be the usual unit vector of the z-axis. Let $\mathbf{f} = \mathbf{k} \times \nabla F$. Compute $\int_C \mathbf{f}$.

Solution. We see that the line integral of $\mathbf{f} = \mathbf{k} \times \nabla F$ on C is the same as the surface integral of ∇F. The argument is the same as in the solution to Problem 10.39. Hence, from the solution of Problem 10.40 we obtain

$$\int_C \mathbf{f} = \int_{\mathcal{S}_E} \nabla F = 2\pi \sum\nolimits_{i \in I_E} c_i, \quad \text{where } I_E = \{\, i \mid \mathbf{a}_i \in E^o \,\}.$$

10.43 Verify Green's theorem for $\int_C (x^2 - xy^3)dx + (y^3 - 2xy)dy$, where C is the square with vertices $(0,\, 0)$, $(2,\, 0)$, $(2,\, 2)$, $(0,\, 2)$.

Solution. Here $P(x,\, y) = x^2 - xy^3$, $Q(x,\, y) = y^3 - 2xy$, and

$$G(x,\, y) = \left(\frac{\partial Q}{\partial x} - \frac{\partial P}{\partial x}\right)(x,\, y) = 3xy^2 - 2x.$$

If E is the interior of C then $\int_E G = \int_0^2 \int_0^2 (3xy^2 - 2x)\, dxdy = 8$. Also,

$$\begin{aligned} \int_C (Pdx + Qdy) &= \int_0^2 x^2\, dx + \int_0^2 (y^3 - 4y)\, dy \\ &\quad + \int_2^0 (x^2 - 8x)\, dx + \int_2^0 y^3\, dy = 8. \end{aligned}$$

Hence Green's theorem is verified for this case.

10.45 Choose $\mathbf{h}(x,\, y) = x\mathbf{j}$ to obtain the area of E as a line integral on the boundary of E. Use this result to find the area of the region bounded by the curve $x = a\cos^3 t$, $y = a\sin^3 t$, $0 \le t \le 2\pi$.

Solution. If $P(x,\, y) = 0$ and $Q(x,\, y) = x$ then $(Q_x - P_y)(x,\, y) = 1$. Hence, if E is an almost regular region in the xy-plane then $\int_{\partial E} xdy = \int_E dxdy$ is the area of E. If E is the region bounded by the curve $x = a\cos^3 t$, $y = a\sin^3 t$, $0 \le t \le 2\pi$, then the area of E is

$$\int_0^{2\pi} 3a^2 \cos^3 t\, \sin^2 t\, \cos t\, dt = 3a^2 \int_0^{2\pi} \cos^4 t \sin^2 t\, dt = 3a^2\pi/8.$$

10.6 STOKES' THEOREM ON MANIFOLDS

10.47 Show that $\nabla \cdot \nabla \times \mathbf{f} = 0$.

Solution. Let $\mathbf{f} = P\mathbf{i} + Q\mathbf{j} + R\mathbf{k}$. Then, with partial derivatives as subscripts,

$$\begin{aligned} \nabla \times \mathbf{f} &= (R_y - Q_z)\mathbf{i} + (P_z - R_x)\mathbf{j} + (Q_x - P_y)\mathbf{k} \text{ and} \\ \nabla \cdot \nabla \times \mathbf{f} &= (R_{yx} - Q_{zx}) + (P_{zy} - R_{xy}) + (Q_{xz} - P_{yz}) = 0. \end{aligned}$$

The last step follows from the interchangeability of the order of differentiation.

10.49 Verify the classical Stokes' theorem, Theorem 10.6.10, for the following cases of E and $\mathbf{f}$, by computing the integrals $\int_{\partial E} \mathbf{f}$ and $\int_E \operatorname{curl} \mathbf{f}$ separately.

1. S is given as $z = (x^2 + y^2)^{1/2}$ and $1 < z < 4$ and $\mathbf{f}(x, y, z) = (z^2, x^2, y^2)$.
2. S is given as $z = x^2 + y^2$ and $z < 4$ and $\mathbf{f}(x, y, z) = (z, xz, x)$.
3. S is given as $x + y + z = 1$, $0 < x$, $0 < y$, $0 < z$, and $\mathbf{f}(x, y, z) = (y, z, x)$.

Solution. We will orient all the surfaces with upward pointing normals $\mathbf{n}$ so that $\mathbf{n} \cdot \mathbf{k} > 0$. The orientation of the boundary is determined by the orientation of the surface. We use the notations in Remarks 9.4.9 for the computations of the surface integrals the notations in Definition 9.4.4 for the computations of the line integrals.

1. We have $\mathbf{g} = \nabla \times \mathbf{f} = 2(y, z, x)$. A parametric representation of S is

$$\mathbf{r}(u, v) = (u, v, (u^2 + v^2)^{1/2}), \ (u, v) \in U,$$

where $U = \{ (u, v) \mid 1 < (u^2 + v^2)^{1/2} < 4 \}$. Hence, with $\rho = (u^2 + v^2)^{1/2}$,

$$H = \mathbf{g} \cdot \mathbf{r}_u \times \mathbf{r}_v = \begin{vmatrix} 2v & 2\rho & 2u \\ 1 & 0 & u/\rho \\ 0 & 1 & v/\rho \end{vmatrix} = 2(-uv/\rho - v + u).$$

To compute the integral on S use the polar coordinates $u = \rho \cos \vartheta$, $v = \rho \sin \vartheta$.

$$\begin{aligned} \int_S \mathbf{g} &= \int_U H(u, v)\, du dv \\ &= 2 \int_1^4 \int_0^{2\pi} \rho(-\cos\vartheta \sin\vartheta - \sin\vartheta + \cos\vartheta)\rho \, d\vartheta d\rho = 0. \end{aligned}$$

The boundary of S consists of two circles

$$\begin{aligned} C_1 &= \{ (x, y, z) \mid x^2 + y^2 = 16, \ z = 4 \} \text{ and} \\ C_2 &= \{ (x, y, z) \mid x^2 + y^2 = 1, \ z = 1 \}. \end{aligned}$$

The upper circle C_1 is oriented in the usual positive direction as $x = 4\cos\vartheta$, $y = 4\sin\vartheta$, $z = 4$, $0 \le \vartheta \le 2\pi$, and the lower circle in the negative direction. Hence

$$\begin{aligned}\int_{C_1} \mathbf{f} &= \int_0^{2\pi} (16(-4\sin\vartheta) + (4\cos\vartheta)^2 4\cos\vartheta)d\vartheta = 0 \text{ and}\\ \int_{C_2} \mathbf{f} &= -\int_0^{2\pi} ((-\sin\vartheta) + (\cos^2\vartheta\cos\vartheta)d\vartheta = 0.\end{aligned}$$

Hence Stokes' theorem is verified in this case.

2. We have $\mathbf{g} = \nabla \times \mathbf{f} = (-x, 2, z)$. A parametric representation of S is

$$\mathbf{r}(u, v) = (u, v, (u^2 + v^2)), \ (u, v) \in U,$$

where $U = \{ (u, v) \mid (u^2 + v^2)^{1/2} < 2 \}$. Hence, with $\rho = (u^2 + v^2)^{1/2}$,

$$H = \mathbf{g} \cdot \mathbf{r}_u \times \mathbf{r}_v = \begin{vmatrix} -u & 2 & \rho^2 \\ 1 & 0 & 2u \\ 0 & 1 & 2v \end{vmatrix} = (2u^2 + \rho^2).$$

To compute the integral on S use the polar coordinates $u = \rho\cos\vartheta$, $v = \rho\sin\vartheta$.

$$\begin{aligned}\int_S \mathbf{g} &= \int_U H(u, v)\, dudv\\ &= \int_0^2 \int_0^{2\pi} (2\rho^2\cos^2\vartheta + \rho^2)\rho\, d\vartheta d\rho\\ &= 4\pi \int_0^2 \rho^3 d\rho = 16\pi.\end{aligned}$$

The boundary of S is the circle

$$C = \{ (x, y, z) \mid x^2 + y^2 = 4, \ z = 4 \}.$$

This circle is oriented in the usual positive direction as $x = 2\cos\vartheta$, $y = 2\sin\vartheta$, $z = 4$, $0 \le \vartheta \le 2\pi$. Hence

$$\int_C \mathbf{f} = \int_0^{2\pi} (4(2\cos\vartheta)(-2\sin\vartheta) + 4(2cos\vartheta)(2\cos\vartheta))d\vartheta = 16\pi.$$

This verifies Stokes' theorem in this case.

3. We have $\mathbf{g} = \nabla \times \mathbf{f} = -(1, 1, 1)$. A parametric representation of S is

$$\mathbf{r}(u, v) = (u, v, (1 - u - v)), \ (u, v) \in U,$$

where $U = \{ (u, v) \mid 0 \leq u, 0 \leq v, u + v \leq 1 \}$. Hence,

$$H = \mathbf{g} \cdot \mathbf{r}_u \times \mathbf{r}_v = \begin{vmatrix} -1 & -1 & -1 \\ 1 & 0 & -1 \\ 0 & 1 & -1 \end{vmatrix} = -3. \text{ Therefore}$$

$$\begin{aligned} \int_S \mathbf{g} &= \int_U H(u, v)\, dudv \\ &= \int_0^1 \int_0^{1-u} (-3)\, dvdu = -3/2. \end{aligned}$$

The oriented boundary C of S is the triangle that joins the vertices starting from $(1, 0, 0)$ to $(0, 1, 0)$ then to $(0, 0, 1)$ and then back to $(1, 0, 0)$. The first segment is represented as $x = 1 - t$, $y = t$, $z = 0$, where $0 \leq t \leq 1$. The other two segments are similar. Hence

$$\begin{aligned} \int_C \mathbf{f} &= \int_0^1 (t, 0, 1-t) \cdot (-1, 1, 0) dt + \int_0^1 (1-t, t, 0) \cdot (0, -1, 1) dt \\ &\quad + \int_0^1 (0, 1-t, t) \cdot (1, 0, -1) dt = -3/2. \end{aligned}$$

This verifies Stokes' theorem in this case.

Printed and bound by CPI Group (UK) Ltd, Croydon, CR0 4YY
23/09/2026
02221785-0002